PRACTICAL BEE KEEPING AND HONEY PRODUCTION

By

D. T. MACFIE

ILLUSTRATED

LONDON

W. H. & L. COLLINGRIDGE LTD.

2-10 TAVISTOCK STREET, COVENT GARDEN, W.C.2

A
FOREWORD

THERE are two ways of endeavouring to awaken and foster an interest in Apiculture, or Bee Keeping. One is to paint a highly coloured word picture or tell a glowing story about handsome profits obtainable from a trifling capital outlay and representing the labour involved in management of a few hives as being purely a pleasant manner of occupying a leisure hour at odd times. In their desire to enlist recruits to the ranks of bee keepers some enthusiasts have adopted some such policy, and first appearances have tended to encourage the opinion that those are tactics which prove their success by results.

We have, in fact, known considerable numbers of people who, after hearing or reading of great weights of honey being secured from one or two hives of bees, have invested a pound or two in starting what they hoped would speedily develop into thriving and lucrative apiaries. Alas, too often the handicaps have proved to be too formidable, and a great many people who have begun with an effervescing enthusiasm have soon dropped their new hobby with feelings of disillusionment and disappointment!

The other way to increase and advance a genuine and lasting interest in this pursuit is to be clear and candid at the outset and all the way through, stating plainly when, where, and by whom bees may be kept with reasonable prospects of success. Then it is well to tell those who are anxious to learn just those details of management which are essential to the welfare of the bees and calculated to increase the quantity and maintain the quality of the honey they produce.

A Foreword

The first few paragraphs of this book will suffice to carry conviction that the author entertained no idea of adopting the unwise policy we have first described. The second plan was undoubtedly made his aim at the outset, and careful study of each chapter in proper sequence will afford ample proof that he faithfully adhered to his purpose right to the end of his task. We have no hesitation in adding that even the most inexperienced novice should feel quite competent to venture upon bee keeping on practical lines when he has assimilated the very sound information here placed at his disposal.

Apiculture has, indeed, made very remarkable strides of progress since the days of our earliest personal acquaintance with bees in the old-fashioned bell hives of plaited straw. Even when we advanced so far as to possess our first wooden hives, they were rather crude affairs by comparison with the types of structures in general use to-day. In fact, we readily confess that, after the lapse of some years since we had the handling of bees, much that we have read of Mr. Macfie's up-to-date information has served to impress us with the realisation that scientific apiculture has left us sadly in the rear.

We do not consider it to be in the nature of bold prophecy to express the opinion that this book will be the means of starting many at bee keeping and will help them to get well away on the road to success.

A. J. MACSELF.

CONTENTS

CHAPTER   PAGE

I.   BEE KEEPING AS A PRACTICAL PROPOSITION   13

II.   HOW BEES WORK -   -   -   -   15

III.   THE MODERN BEEHIVE   -   -   -   22

IV.   FRAMES AND COMB FOUNDATION   -   -   26

V.   OTHER ESSENTIAL APPLIANCES   -   -   32

VI.   THE VARIOUS RACES OF HONEY BEES -   37

VII.   MAKING A START   -   -   -   -   40

VIII.   HANDLING BEES -   -   -   -   -   45

IX.   THE HONEY HARVEST -   -   -   -   52

X.   SWARMING -   -   -   -   -   -   61

XI.   FEEDING, WINTERING, AND SPRING
        MANAGEMENT -   -   -   -   69

XII.   UNITING AND INCREASING STOCKS -   -   79

XIII.   QUEEN REARING -   -   -   -   84

XIV.   EXTRACTING AND HANDLING HONEY   -   90

XV.   OTHER ITEMS OF IMPORTANCE   -   -   97

XVI.   A BEE KEEPER'S CALENDAR -   -   -   106

XVII.   PESTS AND DISEASES   -   -   -   109

XVIII.   MARKETING HONEY   -   -   -   117

INDEX   -   -   -   -   -   121

FULL-PAGE PLATES

	FACING PAGE
SPRING CLEANING THE APIARY - - - -	16
WHERE WATER IS SCARCE - - - - -	17
LATH SHELTER IN USE - - - - -	32
A SHELTERED APIARY - - - - -	33
TAKING A SWARM - - - - - -	64
HIVING A SWARM - - - - - -	65
REMOVING A SUPER - - - - - -	80
QUEEN CELLS AND A MODERN HIVE - - -	81

ILLUSTRATIONS IN TEXT

	PAGE
QUEEN EXCLUDER ZINC - - - - -	19
THE W.B.C. TELESCOPIC HIVE - - - -	23
A FRAME FOR A MODERN TYPE OF HIVE - -	26
A FRAME BOARD OF PRACTICAL DESIGN - -	28
THE WOIBLET SPUR EMBEDDER - - - -	30
A USEFUL TYPE OF SMOKER - - - -	34
A FITTED SECTION CRATE - - - - -	54
A SUPER CLEARER - - - - - -	58
A BEE ESCAPE IN DETAIL - - - - -	60
THE PERFECTION FEEDER - - - - -	70
AN EXTRACTOR FOR THE SMALL BEE KEEPER - -	92
THE SOLAR WAX EXTRACTOR - - - -	94

Chapter one

BEE KEEPING AS A PRACTICAL PROPOSITION

WHEN is bee keeping a practical and profitable proposition? That is a question which is very frequently asked. The answer is not so easily given. Bees can be remunerative by comparison with other agricultural side-lines, and well kept they show a profit far exceeding that of any other. But before handsome returns can be confidently anticipated, there are several factors to be taken into account.

It is useless, for example, to expect an abundant honey flow in large industrial areas or in the suburbs of great cities. Cases are recorded where bees have even been kept on the roofs of city factories, but such colonies will cost more to feed than they will return in the way of honey. Similarly in the case of suburban gardens. Flowers in near-by gardens do not provide the bees with a wide enough field.

Where food is ample, as in clover and heather districts, is where bees will, if properly tended, show a really handsome profit, though the impression, still prevalent, that bee keeping consists merely of putting a colony in a hive and taking a hundred pounds of honey off at the end of the season, must not be fostered. A really strong colony will yield as much as fifty to sixty shillings clear profit in a good year, but a fair average must be counted nearer to twenty or thirty shillings. That, in view of the fact that a start can be made for as little as four or five pounds, must be reckoned a more than usually remunerative return.

There is, of course, one thing which must always be taken into account with any agricultural or horticultural venture in this country, and that is the climate. In a bad season, which in the case of bees means a wet season, they will be able to do nothing more than keep themselves. It must, too, be clearly understood that bees, to produce a handsome surplus, must have proper care and attention. This does not mean that the hobby makes great demands on the bee keeper's time. For nearly half of the year there is nothing to be done, and from the spring onwards an average of fifteen to twenty minutes per hive per week is all that is required. Another point—an apiary takes up little space. A single hive can be accommodated in four square feet and quite a number in an average-size garden.

To sum up, it may be said that bee keeping is not an occupation at which anyone should attempt to make a living, for, as already mentioned, it is much too dependent on climatic conditions, and the vagaries of the weather in this country are not notorious without reason. The countryman, however, even though he have little space at his disposal, will find it a profitable occupation, especially as there is always a demand for good quality honey. In proof of this fact it need only be mentioned that every year many thousands of pounds of foreign honey are imported into this country, much of it inferior, and none of it quite equal to the standard of the home-produced article.

Honey, too, when extracted is more or less imperishable. Carefully stored it will keep for years in perfect condition, with the result that there is no need to unload one's stores on a fallen market during plentiful years.

As a hobby pure and simple bee keeping ranks high, and, despite the wholesome regard with which those uninitiated in the craft are apt to regard the insects, there are few more delightful recreations. Properly handled bees are more or less harmless, and the mind cannot fail to benefit from the study and appreciation of their wonderful social life.

Chapter two

HOW BEES WORK

BEFORE ever a start is made, the would-be bee keeper must understand something of the life and the habits of the honey bee. This does not imply laborious research into the insect's natural history; all that is required is a fair knowledge of how a colony works and lives.

The honey bee (*Apis mellifica*) differs in one important respect from the bumble bees and the various other kinds indigenous to this country. They are social insects, living in large colonies and working for the common good. Most of the other types are solitary or non-gregarious. It is this social habit that makes the honey bee so valuable an insect. By reason of their numbers they are enabled to maintain a winter temperature in the hive which is high enough to ensure their survival in any normal winter. Were the numbers of any one colony to be greatly reduced, it is very doubtful if they could come through unscathed, for the smaller the number of bees in a hive the lower the temperature will be, and the insects are extremely susceptible to cold.

There are three different classes of bee in a colony : the Queen, the worker, and the drone. So distinct are these that they are frequently regarded as three sexes, though this is not correct from the scientific standpoint. The Queen is the only fully developed female. She is the head of the colony and the mother of every bee in the hive. Hers is the responsibility of carrying on the race, and this she is capable of doing for four or five years, laying eggs in the height of the breeding season at the stupendous rate of two, or even three, thousand a day. It is on the prolificacy and the temperament of the Queen that the welfare of the hive depends, and for reasons that will be detailed later on too much importance cannot be attached to seeing that each hive is headed by a young, prolific, and healthy mother.

In appearance, the Queen is quite distinct from either drones or workers. She is larger than either, and of more graceful build, long in the body and with wings which, by comparison with the abdomen, appear comparatively short. Like the worker, she has a sting, but rarely, if ever, is it used except on others of her own sex. To persuade a Queen bee to sting a human being is wellnigh impossible.

The worker bee, which constitutes by far the greatest proportion of the population of the hive, is the smallest of the three classes. Actually, the worker is simply an imperfectly developed female, and it is on her that the entire labour of the hive devolves. So great is this social desire for work that she literally toils herself to death. Shirkers are unknown, and it is computed that the average life of these selfless labourers is no more than four or five weeks during the honey season. Then, with their wings worn and frayed to such an extent that they can no longer fly, they may be found crawling towards the hive they cannot reach before night brings a drop in temperature, which inevitably entails their death.

There are multitudinous tasks to be performed in a bee community, and the division of labour in a hive is something at which to marvel. First of all, there are the larvæ in the brood combs to be attended to. This is the first task for newly hatched workers. The Queen, too, must have her retinue of attendants; others must be menials clearing debris from the hive, for there is no more cleanly insect than the bee. Ventilation must be

SPRING CLEANING THE APIARY.   THE BEES ARE MOVED TO A FRESH
HIVE, AFTER WHICH THE ONE IN WHICH THEY HAVE SPENT THE
WINTER IS THOROUGHLY CLEANSED.

WHERE WATER SUP-
PLIES MAY NOT BE
PURE IT IS WISE TO
PROVIDE DRINKING
FOUNTAINS AS SHOWN
IN THE FOREGROUND.
IMPURE WATER IS
FREQUENTLY THE
CAUSE OF DISEASE.

maintained, and so the task of fanning their wings at the entrance occupies yet more; while all must be prepared to give their life, if need be, in defence of the home. Most important of all of the indoor tasks, there is wax to be made and cells to be built, the latter with a mathematical accuracy that man himself cannot excel.

First and foremost amongst all the workers' tasks, however, is that of gathering nectar and pollen from the flowers, and this is the portion of most as soon as they are strong enough to fly afield. For this purpose they are furnished with a sac or honey stomach, in which the nectar is stored, while on the hind legs are little baskets for conveyance of pollen. The worker bee, as is well known, is equipped with a sting which is freely used when occasion demands, but the act of stinging almost invariably results in the death of the bee, for it is seldom able to withdraw the sting with its circular barbs.

The drone is the male bee, and a big, bulky fellow. His purpose in life is only one, and that is the fertilisation of the Queen. That apart, he takes no part in the work of the colony, but lazes the summer through, not one in a colony, but three or four hundred in each. His is the lotus eater's life at first appearance, but his fate is not an enviable one. One or two in every three or four hundred fulfil their lives' purpose and mate with Queens. This act alone entails the death of those favoured and the days of those remaining are numbered, for with the autumn they are ruthlessly driven from the hive and left to perish from cold. Defend themselves they cannot, for Nature has not seen fit to equip them with a sting. Nor are they capable of fending for themselves in the way of gathering food.

Some reason for the number of drones in a hive is to be found in the fact that the whole life of the colony centres around the Queen. A suitable mate for her must be ensured, but from the point of view of honey production it is a mistake to have too many drones, for they are merely a drain on the honey resources. The aim must therefore be to keep their numbers down to the lowest possible minimum. This can be achieved in several ways, as will be detailed in later chapters.

The Queen bee may live for four or five seasons, but it is seldom that she is allowed to do so in Nature, and even more seldom where bees are kept on up-to-date lines. The one mating flight fertilises her for life and she lays fertile or infertile eggs according to the size of the cells in which they are deposited. The former produce workers and the latter drones, the infertile eggs being laid in cells that are rather larger than those which will produce workers. New Queens are raised from the same eggs as workers, but their upbringing is a totally different matter. In the first place, they are laid in special cells, three times the size of the worker cells and usually hung from the edge of the comb. The larvæ, too, are fed on very special food called Royal Jelly. This is made from pollen and honey, and is partially digested by the bees themselves before being fed to the larvæ. All the larvæ in the hive are fed on a coarser form of this same food, but for only three days. The grubs in the queen cells are given lavish supplies throughout the whole period of their development, and it is this royal treatment which enables them to grow into perfect bees.

It is a curious fact that, although the Queen is much the largest of all bees, she emerges from the cell in less time than do either workers or drones. Sixteen days from the laying of the egg is the average in her case as against twenty-two for the worker and twenty-five for a drone, and right at the very start of her life the young Queen bee emerging from her cell faces its biggest issue, for it is very, very rare to find two Queens

18

in one hive. As more than one queen cell will have been built, there may be others already hatched out or on the point of so doing in addition to the possibility of the old Queen still being in the hive. The latter will, however, in all probability have headed a swarm as soon as the first of the queen cells were capped over.

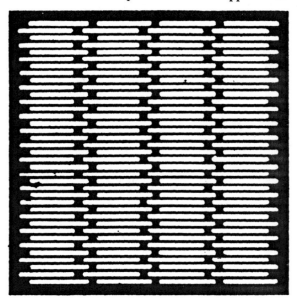

QUEEN EXCLUDER ZINC

The slots are wide enough to allow free passage for the worker bees, but they prevent the Queen passing into the supers and laying eggs in shallow frames or sections.

Straightway the new-comer makes a search for possible rivals. If they are present, a battle is fought for sovereignty of the hive, the loser's fate being death, the victor's a life of ceaseless egg laying. Just to ensure against further distractions from this all-important task, the victorious Queen's next move is to examine the hive minutely for further queen cells, whose occupants are ruthlessly destroyed.

Within a week or ten days the young virgin Queen will undertake her nuptial flight. From that day onwards she will not leave the hive, unless in the early summer of the following year, when she may head a swarm to seek fresh quarters. This is a habit which invariably puzzles the novice. Actually, it is the bees' natural method of increasing colonies when a hive becomes overcrowded and takes place as soon as the queen cells, which are to produce the new monarch, have been capped over. Once they are assured that their hive will not be left without a mother, the Queen and a fair proportion of her subjects set forth to find new quarters, leaving their home to those which remain and the larvæ in the brood.

Hollow trees are the invariable home of wild honey bees. There in dry shelter they build parallel combs hanging from and firmly attached to the ceiling of their home. If bees are kept in the old-fashioned skeps, their procedure is identically the same. The skep, within a comparatively short space of time, is completely filled with combs just far enough apart to allow the workers to pass. Keeping bees under these conditions had many obvious disadvantages. In the first place, eggs, larvæ, pollen, and honey were all on the same combs, and to gather the honey harvest only two courses were open to the bee keeper: one, to kill off all the inhabitants of the skep by means of sulphur fumes, and two, to frighten or drive the bees to another empty skep. As a result much time and thought were devoted to evolving a type of hive in which the bees' energies could be to some extent controlled. The result was the movable frame hive, which, with only slight structural alterations, is in universal use to-day and that after a period of fifty years and more. The great principle of the movable frame hive is to confine the Queen to one chamber, but to allow the workers free access to all. As a result, breeding is completely limited to this one lower section, or two sections under certain circumstances, while the upper ones are reserved for the main honey stores. The actual construction and sections of a modern hive are more fully explained in a later chapter.

The Queen is prevented from leaving the brood chamber by means of a simple but ingenious device known as a queen excluder. This is usually a sheet of zinc in which are cut slots wide enough to allow passage for the workers, but not the larger Queen. Another type of excluder consists of a wooden frame with cross bars of stout galvanised wire, cross braced for strength and set so close together that a Queen bee cannot pass between them though there is just space for a worker.

Another point is that, instead of allowing the bees to attach their combs to ceiling or walls, movable frames hanging from lugs are placed in position for them, all ready provided with wax foundation on which is impressed the shape of the cells. In this way the bees are saved a colossal amount of labour, for when placed in a hive they have only to draw out the cell walls, and a certain amount of surplus wax is provided to assist them in this work. Thus they are enabled to devote the maximum energy to honey production.

It must, of course, be realised that in theory the only honey which can be removed is that which is superfluous above the bees' own requirements. The store they labour to hoard is there to feed them during the winter months. In actual practice it is seldom that the bees are allowed to retain enough for their subsistence, but the loss is made good to them by means of feeding with syrup and candy.

THE MODERN
BEEHIVE

BEE keeping as a science practically dates from the introduction of the modern bar frame hive. As explained in the preceding chapter, the Queen in a bar frame hive is prevented from leaving the one lower chamber, and the bees can be provided with foundation wax on which to commence their labours. But these are only two of the advantages it possesses over the primitive skep. Another great point is that the bee keeper is enabled to examine the combs at will and so ascertain the health and condition of the bees and brood. If, as sometimes happens, the Queen is lost and there are no young larvæ or eggs which the workers can transfer to queen cells, the colony is doomed to extinction unless a new Queen be introduced. This in a skep is impossible, as it also is to verify the fact that the Queen is lost. Her decease is soon made apparent in a bar frame hive by the presence of laying workers. These are larvæ which have been transferred to queen cells on the death of the monarch, but too late. To develop into perfect Queens the grubs must receive the royal treatment before they are three days old. If it is delayed beyond that time, they develop into bees which can lay but which cannot mate with the drones. As a result they lay only infertile eggs, and these, of course, produce only drones. Their presence in a hive is a certain sign of queenlessness.

Again, it is essential to find out just what stores of honey each colony has laid by for the winter. Like queenlessness, the state of the food stores can only be assessed with accuracy in a bar frame hive.

There are numerous types of hives in use, which differ only in structural detail, but in this country all are manufactured to accommodate the standard British frames. The hive which has undoubtedly the widest appeal is the W.B.C. This is of the usual tiering type.

THE W.B.C. TELESCOPIC HIVE

One side has been removed to show details of the interior arrangement. At the bottom is the brood box. Above that a rack of shallow frames, and on top a crate of shallow sections.

That means simply that the various sections or chambers are tiered one above the other, and in theory there is no limit to the number of upper or honey chambers which can be used. The bottom chamber, or brood body, of the hive rests on the floorboard with a bee space of half an inch below the front edge. It is built to accommodate ten standard size brood frames, which hang at right angles to the entrance. Immediately above this brood body and on top of the frames the queen excluder is placed.

The next storey or " super," as it is technically termed, accommodates a rack of shallow frames used for obtaining extracted honey or it may be a section rack if it is intended to secure honey in the comb. Above this again may be other supers or section racks according to the requirements of the moment.

Outside this hive proper is an outer casing built up of successive sections, or lifts, fitting closely one on top of the other. These can be added to as necessary so as to provide extra height for the supers or section racks within. There is a clear space of an inch or so between the inner hive and this outer casing. This is a great help in preventing changes of temperature, and does much also to prevent the bees suffering from excessive cold in winter and from over-warm temperatures in summer.

On top comes the roof, ridged or tilted to shoot off rain. Above the top super, quilts are laid to keep the temperature up, and just in case of some bees accidentally finding their way through the quilts, a bee escape is fitted in the roof. This is in the form of a perforated cone, which provides ready exit but prevents the bees re-entering. Incidentally, the fitting of a cone escape will also serve to provide necessary ventilation.

Although from the foregoing description it may sound a simple job for any handyman to knock a hive together, it is inadvisable for anyone but a skilled craftsman to attempt to do so. The job is one for the cabinet-maker and not for the rough-and-ready carpenter, for if there is the least error in construction, trouble will result. Every part must fit with absolute accuracy, or the bees will set to work with propolis or brace comb. This is simply because they will not tolerate spaces larger or smaller than that required for their own passage. A " bee space " is approximately one-quarter of an inch. Any space less than this is liable to be filled with propolis—a sticky substance which is also known as bee glue, and is obtained from the resinous substances exuded from tree buds. If any spaces are left much exceeding a quarter of an inch they will be filled with comb. In fact, it is necessary to build with a maximum allowable error of three thirty-seconds either way, which is rather beyond the skill of the average handyman.

It is, moreover, a job which cannot be considered worth while in view of the fact that hives can nowadays be purchased at very reasonable prices. They may be obtained fully assembled and fitted or in the flat. Assembling in this latter case is a very simple matter, as practically all joints are dovetailed or grooved and tongued and the parts are very accurately turned out.

Deal was for long the favourite wood, but preference nowadays should be shown for Western red cedar for the outer lifts. This Empire timber is absolutely impervious to rot of any kind and requires no painting, though it is customary to paint it when used for hives because bees like to be able to see their hives clearly at a distance. The fact that most of the trade experts have already adopted Western red cedar to the exclusion of other timbers is in itself sufficient commendation.

Chapter four

FRAMES AND COMB
FOUNDATION

THE frames used in the modern hive are of three types—brood frames, shallow frames, and sections. The former are housed in the lower brood body of the hive, and are, of course, those in which the young larvæ are raised. Shallow frames are identical in every respect, except that they are not so

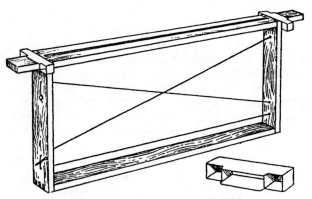

A FRAME FOR A MODERN TYPE OF HIVE

The frame has been wired ready to receive a sheet of foundation wax and metal spacing.

deep, and are used only in the supers for the storing of honey which is to be extracted. Sections are the small white-wood boxes used in the production of comb honey. All are made in strictly standard sizes, and as they can be purchased very cheaply, it is inadvisable for the bee keeper to attempt to make them himself. As with the hive itself, it is essential that their measurements should be exact within a fraction of an inch, for they must hang at just the right distance apart, and it must be possible to lift each one from the hive without tearing asunder any attachment to another comb or to the hive walls.

Brood and shallow frames consist simply of four pieces of wood, which are jointed together. The top bar, including the lugs, measures 17 inches, the bottom bar 14 inches, and the side bar 8½ inches. In shallow frames the side bar measures only 5½ inches. The section size in general use measures 4¼ × 4¼ × 1 $^{15}/_{16}$ inches. This holds approximately one pound of honey. Sections are made in one piece and supplied in the flat, grooved for folding, and fastened with a dovetail joint. It is a very simple matter to assemble them in readiness for fitting comb foundation. It is as well, however, to damp the joints with water before folding, as unless this is done, there is a risk of breakages.

Assembling frames is also a simple task, as all the ends are jointed; but it is advisable, for the sake of additional security and stability, to supplement the joints with thin nails. Those known as panel pins are most suitable. To ensure that the frames hang at just the right distance apart in the hive, metal spacers are fitted over the lugs. These too are very cheaply purchased.

The fitting of foundation to the frames and sections is a job which requires a little practice before it can be accomplished with any celerity, but it is really quite a simple matter and one which every would-be bee keeper should set himself to master from the outset.

There is no doubt that the introduction of foundation wax has done more than any other thing to put bee keeping on a really paying basis. It is simply a sheet of pure beeswax on which is stamped out by machinery the form of the cell bases. These are of exactly the same form and measurements as in the natural comb, and all that the workers have to do is to draw out the cell walls. Some idea of the labour saved can be gained when it is realised that the bees themselves, to make one pound of wax, must consume something between ten and fifteen pounds of honey. There is another point: the use of foundation helps to decide the type of cells

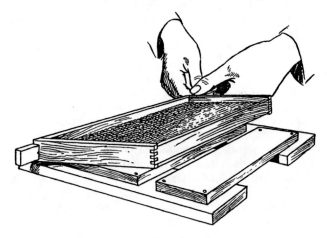

A FRAME BOARD OF PRACTICAL DESIGN

This is for use when fixing wax foundation. The raised boards are of two sizes, one to fit brood frames and the other for shallows, and are just under half the depth of the frames, so that the sheets of wax lie flat upon them.

that are built. In the brood chamber worker foundation is used, i.e. foundation stamped with the shape of the small worker cells. This reduces the number of drone cells built. On the other hand, in the supers, drone foundation is used because the cells are bigger and therefore more economical from the honey-storing standpoint. There is no fear of their being used for

28

drone breeding, as the Queen is excluded from these supers.

The foundation is made in sheets of varying thicknesses. That for the brood frames is thick and of yellow wax. Super foundation is thinner and of lighter-coloured wax. A heavy foundation must always be used in the brood frames, and it is usually advisable to make use of it also in shallow frames which are to be extracted, for with the thicker sheets there is less risk of breakages. For sections the thinner super foundation is preferable. There is no need to place any drone foundation in the brood chamber, as the bees themselves will always make the necessary alteration to include the requisite number.

The foundation is sold in sheets by the pound, the thickest running about eight or ten to the pound, and in sizes to fit sections or brood frames. Full sheets should be used in every case. The cost is actually trifling as compared with the labour saved, and there should never be any thought of using only half sheets or starters, which are narrow strips of foundation placed under the top bar.

Inserting the foundation in sections is a very simple matter, for all that is necessary is to fit it into the grooves cut in readiness. With frames it is fitted into a saw-cut in the top bar, but first of all the frames themselves must be wired for additional security. Unless this is done, the combs are liable to fall out, especially during warm weather and when heavy with honey.

There are many methods of wiring frames, and it is of little account which is adopted so long as the wire is stretched really taut. A typical one is shown in the accompanying illustration. Three holes are bored in each of the side bars. Short panel pins or tacks should also be driven half-way home near the single holes but on the outside of the bars. Tinned wire is then threaded through the holes, and the ends secured to the tacks, after which the loops should be tightened, and the tacks driven home. The ideal is to have the wire so taut that it emits a musical note when touched, but care must be taken to see that the bars are not drawn out of parallel.

Both brood and shallow frames are made with either

THE WOIBLET SPUR EMBEDDER

The toothed cog wheel is heated and run along the wires, which sink into the melted wax. The grooves cut in each cog are shown in the inset.

a saw-cut running almost from end to end of the top bar, or they are grooved on the underside to hold the foundation. This latter type, which has also a small wedge that is driven home to hold the foundation firmly in position, is undoubtedly the easier for beginners to use.

The next step is to embed the wire in the wax. This

is quite easily done with the aid of a spur embedder, an ingenious little tool with a toothed cog wheel, each cog having a groove in which the wire is held. This is heated and then run along the wires. The heat melts the wax just around the wire and as a result the latter sinks into it and is firmly held. The only point to watch is that the tool is not too hot or it may cut through the sheets.

A frame block is essential for this work. This is a piece of wood which exactly fits the inside of the frames, but is just under half their depth. On this the frames with the foundation already fitted in the grooves can be laid. The block will support the foundation and keep it perfectly flat, thus reducing the possibility of accidents. Wiring is not necessary in the case of sections.

If the beginner is at all doubtful of his ability to wire and fit foundation, the difficulty is readily overcome by purchasing frames already fitted. These are not expensive, and will prove a boon where time is limited.

CHAPTER	OTHER ESSENTIAL
FIVE	APPLIANCES
✽	

THERE are various other items with which the
bee keeper must provide himself before he can
consider his outfit complete or even adequate.
First of all, a veil will certainly be required to protect
the face and neck from stings. It is true that there
are some stocks which can be handled with impunity
and without the aid of intimidants or protection, but
the novice is ill-advised to run any risks, and until perfect confidence is acquired, a veil should certainly be
worn when opening the hive.

Almost any kind of fine netting is suitable for the
purpose, but for preference black should be chosen, as
it is more easily seen through. A piece about two feet
by one is amply sufficient. The ends are first sewn
together, after which a piece of elastic is sewn on one
edge. This will grip the stiff-brimmed hat, which must
be worn under the veil to hold it clear of the face.
Elastic can also be sewn round the other edge to hold
it closely round the collar or it can be worn tucked in
under the jacket.

Gloves as a rule are not recommended, as they are
apt to render the operator clumsy, and nothing is so
certain to infuriate the bees as jerky or clumsy movements. They do, however, give the beginner a sense
of confidence, and this is ample compensation for any
clumsiness their use entails. Proper bee gloves with
long gauntlets or rubber gloves are the kind to use.

A smoker is an absolute essential, for there are few
stocks which can be examined without the use of an
intimidant. In view of the price, it is seldom worth

A LATH SHELTER AS SHOWN PROTECTS THE HIVES FROM DIRECT SUN-SHINE AND REDUCES THE RISK OF OVERHEATING DURING THE SUMMER MONTHS

AN APIARY IN A SHEL-
TERED CORNER. SUPERS
ARE IN USE ON MOST OF
THE HIVES, A SURE IN-
DICATION OF STRONG
HEALTHY COLONIES.

while attempting to improvise one. A purchased model, which is fitted with a wire trap in the nozzle to guard against particles of lighted fuel being blown into the hive, is infinitely better. The type in general use consists of a fuel box mounted on bellows and a detachable nozzle. Brown paper, rags, or dry touchwood provide suitable fuel. Whichever is used, it should be placed in the fuel box lighted end downwards. Then, providing the smoker is stood nozzle end upwards when not in use, it will remain alight, thanks to the draught created through the nozzle.

An alternative to the smoker, which is actually preferable for some operations, is a carbolic cloth. This is simply a sheet of calico or muslin about two feet by one and a half, which is soaked in a solution of carbolic. If kept in a tightly closed tin box it will retain its effectiveness for quite a long period.

Feeders will also be required. It is usually wise to have two. One should be of the graduated type for spring stimulation: this is simply an inverted bottle mounted on a screw cap and stage. By turning the bottle the supply of syrup can be regulated to any number between one and nine holes, as indicated by the figures on the stage. In addition, a rapid feeder, which is used chiefly in the autumn when stocks must be fed up quickly so that they may store the syrup for winter use, will be required. The round tin type is thoroughly satisfactory. This is simply a round tin box with a movable lid. The bottom is flanged to provide bee space between the frames and the feeder, and the bees have access into it through a round hole in the bottom. The principle of this feeder is very simple. A wooden float surrounds the funnel into which the bees enter. When the syrup is poured into the tin the float rises, and it is from this that the bees feed. As a rule this feeder is fitted with a glass lid through which

c

the bees can be observed. More detailed information on the use of feeders will be found in a later chapter.

Where extracted honey is worked for, uncapping knives and an extractor will also be necessary, though in many cases it is possible to borrow these aids. Actually, it is possible to use a sharp carving knife to remove the cappings on shallow frames, but the novice

A USEFUL TYPE OF SMOKER

Rags, brown paper, or touchwood can all be used as fuel, and if stood nozzle end upward when not in use, the smoker will remain alight.

is well advised to make use of a proper uncapping knife, as with it there is less danger of damaging the combs. The Bingham and the W.B.C. are the two most popular patterns. The latter resembles a carving knife, but has a turned-up point, thus making it easier to uncap cells which are below the level of the frame containing the comb. The Bingham is a very broad knife with bevelled edges and a trowel handle. Both are eminently satisfactory.

34

❊ *Other Essential Appliances*

The honey extractor must be numbered with wax foundation as amongst the most indispensable aids to bee keeping, for it is only by its use that it is possible to extract the honey without injuring the combs which can later be given back to the bees to refill. Extractors are made in many sizes, and some are of quite elaborate mechanism. The principle in every case, however, is essentially simple.

Briefly, the extractor consists of a round metal cylinder containing two or more cages fixed around a vertical spindle. An uncapped comb of honey is placed in each of the cages, which are then revolved at rapid speed by means of a geared handle. The honey is thrown from the cells by centrifugal motion against the sides of the extractor. As soon as one side of the comb is completed they are reversed in the cages and the operation repeated. The honey in the extractor is drawn off by means of a tap. Quite small but perfectly efficient extractors can be obtained for as little as thirty shillings.

Extractors are unfortunately useless when dealing with heather honey. On account of its much greater density, it cannot be thrown out of the combs. Instead, a press must be used. This, of course, means sacrificing the combs, but the wax itself is saved. There are several makes on the market, all working on almost identical principles, or, if only a very small quantity has to be dealt with, makeshift can be made with a potato masher.

A ripener is another useful though not essential appliance. This is simply a tall, churn-like receptacle, made in two sections. The upper portion is fitted with strainers through which the honey is run into the ripener. It is drawn off from the lower portion by means of a tap. Ripeners are required chiefly for honey which has not been capped before extraction. This cannot be offered for sale unless it is first of all ripened in a warm temperature.

Like ripeners, wax extractors are not essential in a small apiary, but where any quantity of old comb and cappings have to be dealt with, they are a great aid. They are used to render into pure wax cappings, discarded combs, or any odd pieces of comb foundation collected during the season. A simple method of rendering wax without the use of extractors is detailed in a later chapter. Where it is deemed necessary to purchase one, the best type is undoubtedly that operated by steam. This consists of a boiler, above which rests a perforated basket for the comb. Underneath the basket is a tray with an outlet pipe. When the water boils, the steam melts the wax, which pours through into the tray and then to the outlet, where it can be caught in a greased basin. The refuse remains in the basket.

The Solar wax extractor is even simpler to use, but does not, like the steam extractor, render the wax perfectly pure. This is simply a box with a glass lid fitted with a tray and trough. The extractor is fitted with legs at the tray end to keep it on a slope. The wax to be purified is thrown into the tray. The extractor is then placed in the direct rays of the sun. As a result, the wax is melted and runs down into the trough, leaving most of the impurities behind.

Chapter six

THE VARIOUS RACES
OF HONEY BEES

ALTHOUGH hybrid strains are very much favoured by bee keepers to-day, there are several pure varieties which can still be obtained, among which the native brown or black bee is in far the greatest proportion. Nowadays, it is very much crossed and interbred with Italians, Carniolans, and other eastern races. A really pure strain of this variety is, however, still difficult to beat, for these native bees are extremely hardy and well adapted to our fickle climate. The main thing is to procure a really good strain. With experience it will be possible to breed one's own strain, selecting each year the best Queen and breeding only from her. This process of selection carried out over a number of years should produce good stocks of honey gatherers, far superior to the common strains. The main thing is to obtain Queens that are prolific in the spring and with offspring that are not too irritable. As a rule the blacks are quite docile, but some strains are inclined to be a bit vicious. For comb honey production they are strongly recommended, for they are all excellent builders, and their cappings are white and thick. In this respect they are superior to nearly all the foreign bees.

Among the pure strains the Italian or Ligurian comes next in popularity. This is a very handsome bee, and one that is easily recognised by the three yellow bands on the abdomen. Although Italian Queens are much more prolific than the blacks, and the bees themselves are earlier and later workers, they do not, except in mild districts, give better results than a good strain of blacks—probably through an undue mortality amongst the flying bees, as they do not seem quite robust enough for our rather uncertain climate. There are other points which must be laid against them. On account of their very prolific Queens they are rather given to swarming and robbing. Again, they are poor comb builders, and quite unsuitable for the production of really high-class section honey. This is due to the fact that their cappings are placed close down upon the honey, as distinct from the black bee and the Carniolans, which leave an air space. One of their greatest attractions is that they have a very quiet and gentle disposition. Properly handled they can be examined without the use of any intimidants, though this is a risk the novice is not advised to take.

Hybrid strains of Italians and native blacks are very variable in temperament, although some of the best working stocks result from this cross. Some are just as quiet as pure Italians, but others again are so vicious as to be quite unmanageable. Where stocks of this nature are met with, the only remedy is to re-queen from another stock of a more tractable nature.

The Carniolan bee is even more amiable than the Italian, and must be numbered amongst the most suitable varieties for beginners. In appearance they resemble the native black at first sight, but differ from it in having white bands on the lower segments of the abdomen. Like the blacks they are first-class bees for comb honey production, for their cappings are thick and of snowy whiteness. Moreover, they gather very little propolis. The one objection to this variety is that it has an abnormal propensity for swarming. It is quite a common thing, where swarming is allowed to continue unchecked, for a colony to throw off so many casts that the parent stock is left quite worthless. So long as

38

swarming is duly controlled, however, this is not an insuperable objection.

The same fault must be laid against the Dutch bee. In Holland swarming is encouraged, and as a result it is almost impossible to prevent them from so doing when kept under the methods practised in this country. Otherwise they have no faults, being good-tempered, and splendid comb builders and cappers. Their introduction into this country was chiefly in the hope of checking the Isle-of-Wight disease.

Caucasian bees have a reputation for gentleness which does not always appear to be justified. However, it is probable that much depends upon the strain, and it is possible that many so-called pure Caucasian stocks are in reality hybrid. These bees resemble blacks, are very prolific, and make fine white cappings to their cells, but they build an excessive amount of brace comb, which often makes it difficult to remove frames for examination.

There are many other varieties, such as the Cyprians, Egyptians, Syrians, etc., but generally speaking they are seldom met with and need not be taken into account. Not all of them are suitable to our climate, and many of the strains which have been tried have proved extremely wicked in temperament. The beginner is best advised to commence with a really good hybrid strain or the native black. Either, procured from a reliable source, can be thoroughly depended on. Pure Italians might be ranked third, especially in mild districts, and the Carniolans fourth, but it must be borne in mind that there is little chance of keeping any pure race perfectly true for long in a country so thickly stocked with bees.

MAKING A START

AS may be concluded from the foregoing chapters, the beginner who makes a start with a carefully chosen and healthy stock of bees, together with a complete set of the necessary appliances, will possess decided advantages over the man who has merely procured just any stock of bees and a few odds and ends in the way of equipment. One thing is certain. It is never advisable to purchase old apiaries privately unless one has the benefit of the advice of a really experienced man. It is true that such stocks are often offered at extremely low prices, but they may prove costly in the long run.

Disease is the danger. Should any of the stocks acquired be infected, or even if appliances are obtained from an apiary containing disease, there is a grave risk of losing the entire stock. There is no doubt that all appliances, with the possible exception of extractors and honey strainers, should be purchased new or made from fresh clean material, while the brood combs of second-hand stocks should be carefully inspected by an experienced bee keeper for signs of foul brood. This is a disease which not one novice in a hundred can recognise at sight.

It is far preferable to purchase a complete outfit from a reputable dealer. No dealer of any standing will jeopardise his good name by sending out anything but strong and healthy colonies. Complete outfits, including all the necessary appliances, are now offered by leading specialists for as little as five pounds, with an additional two or three pounds for the bees themselves.

Such outfits will include a first-class modern hive, complete with supers containing both shallow frames and sections, wired and fitted with foundation wax, a complete brood body with ten frames containing eggs, grubs, and larvæ, and such additional appliances as a smoker, feeder, super clearer, etc. Such a stock will produce a fair surplus in its first season.

A little economy can be effected by purchasing only a nucleus stock. This will probably mean four frames of bees instead of ten, but it is doubtful if the saving is really worth while, for so small a colony must be given some time to increase before it will produce a surplus. If hives and other appliances are made at home, it is possible to purchase a stock covering as many frames as may be desired, but if all appliances are to be bought, it is undoubtedly wisest to procure everything from the same source. The hive itself should always be procured some little time in advance, in order that it can be fitted up in readiness for the bees' arrival. Moreover, it is a good plan for the novice to familiarise himself with all its fittings before the bees are actually in residence.

The best time to procure stocks is undoubtedly in April or early May, for by then the bees will be flying fairly freely and will soon settle down to steady work. Much will depend upon the season, however, and frequently dealers are not prepared to deliver until the end of May or even early June.

If swarms are preferred, they should be secured, if possible, by the end of May. This is not always practicable, but early June should be regarded as the latest date. Later swarms are small and have not the energy of the first spring ones, though they will with care develop into good stocks. No bees work with quite the same energy and vigour as a natural swarm, and a good one should weigh something between four and six pounds.

Hiving a swarm is quite a simple matter, but it must be done in the correct manner. The hive must of course be fitted out in readiness, complete with brood frames and a super wired and fitted with foundation, and the operation itself should not be attempted until the evening, preferably about dusk. Swarms which are hived in the early afternoon or in the morning may leave the hive the same day.

First of all, boards or hive covers should be placed against the alighting board of the hive and running to the ground. These should be covered with a clean sheet and the entrance of the hive opened to its fullest extent. Now take the skep or the box in which the bees are accommodated, and with a sharp jerk throw them on to the board as close to the entrance as possible. Their natural instinct is to run uphill, and it will not be many minutes before they are all safely inside. Although it is quite likely that the novice may miss her, it is a good plan to watch the bees closely after they are thrown on to the board and endeavour to make sure that the Queen actually enters the hive.

Another method with small swarms is to remove six of the frames in the brood body and throw the bees directly into the hive, after which the frames can be replaced. This is not possible, however, with large swarms, and it is really not a method for the beginner. At best it is simply a time-saver for the experienced bee keeper.

The novice need have little fear of being stung when handling swarms. A veil should certainly be worn just in case of accident, but there is really little risk, for the bees, before leaving their old quarters, will have gorged themselves with honey, and in this state they are seldom dangerous.

As soon as the bees are safely housed, all the quilts, save the topmost one, should be removed, for swarming bees generate a high temperature, and unless this is done there is a risk that the foundation may break down. After an interval of twenty-four hours all the quilts may be replaced.

A freshly hived swarm has, of course, no food stores to which the bees can repair, and in the event of a period of unfavourable weather immediately after hiving it will be necessary to feed them with syrup. This eventuality should not, however, arise, as with quite reasonable conditions they are perfectly capable of meeting their own demands.

Stocks purchased on brood frames do not require any attention in this respect, as they arrive complete with their own store of honey and pollen. Nor is it necessary in their case to remove any of the quilts.

As a rule the queen excluder zinc is placed in position above the frames as soon as the stock or swarm is hived. Providing there is still room for the Queen to lay eggs on the lower brood frames, however, it is sometimes advisable to delay so doing for ten days or a fortnight. There is little likelihood of her ascending to the supers unless the lower brood frames are completely filled, but the workers will do so, and having once become accustomed to using the upper portions of the hive, the presence of the excluder zinc will not deter them from returning.

With stocks which cover only four frames, dummy boards must be used to concentrate the bees in the centre of the hive. These are simply pieces of wood exactly the same size as the brood frames themselves and made with lugs to hang in identical manner. The four frames should be placed in the centre of the hive with a dummy board on each side. As the colony increases in strength the dummies can be moved outwards to allow the bees access to further frames until they occupy the full quota of ten.

The best position for the hive is a fairly sheltered one which is screened from prevailing winds, and which allows of its facing south or south-east. There must, however, be room to approach and work from the back. If the bee keeper attempts to do so from the front the flight of the bees will be interfered with. Another point, the approach should be perfectly open. Hedges, walls, or other obstructions will impede the bees if at all close. A spirit level must be used to ensure that it stands exactly level, for unless it does so, perfectly straight combs will not be built.

Another point to observe is that frames should always hang at right angles to the front of the hive, not parallel to it. If frames are removed for inspection they should always be replaced in the same relative order. When new frames fitted with unfilled foundation are introduced it usually pays to place them in the centre of the hive where the bees are most busily at work.

CHAPTER	HANDLING
EIGHT	BEES

T HERE is no greater mistake than that of opening up hives needlessly, for although it can be done without seeming annoyance to the bees it undoubtedly checks their day's work, with a subsequent loss of honey production. Regular inspection to ascertain the strength of a colony and their health is essential. That apart, the insects should be left as much as possible to their own devices. Bees which have their brood comb constantly pulled about will never make really good progress, and there is also a grave danger of the Queen being killed, either by accident in the moving of the frames, or even by the bees themselves when they cluster over and around to protect her. The golden rule is to examine only when essential, and then to get the job completed in the shortest possible time.

The one great thing essential, if bees are to be handled successfully, is confidence. There is really no need to be nervous, for the beginner can make sure first that he is perfectly protected from possible attack. In any case, it is most unlikely that the bees will show resentment, providing examination is carried out in the correct manner and that all the operations are performed in a quiet, deliberate manner. Bees themselves are very nervous little creatures, and are always liable to resent quick, sudden movements or jarring of the hives or frames. Even after they have been subdued, it is possible for a clumsy or careless operator to upset them. Nothing will do so more quickly than if some bees are crushed in handling the frames. This is always liable to incite them to sudden fury. Another point—they have a very decided objection to human breath, and care must therefore be taken not to breathe directly upon them.

The breaking of comb containing honey is another frequent cause of trouble. The bees apparently think they are being robbed; at any rate they usually attack in force.

When trouble does arise—and it will, occasionally, no matter how careful one may be—there is only one thing to do, cover up the hive and go away. It is probable that in half an hour the bees will be quite tractable once more.

Before ever the hive is opened, everything which is likely to be required should be gathered together and placed close at hand on a box or a small portable table. The smoker should first of all be filled, lit, and tested; some spare fuel should also be laid at hand, for there is nothing more disconcerting than to have it suddenly go out while the hive is still open and it may yet be required. Once alight a few puffs from the bellows will produce an ample volume of smoke, and if stood on end, funnel upwards, it will draw in just the same manner as a chimney. If a carbolic cloth is preferred for subduing the bees, this should be inspected to see that it is in good condition. Other useful aids are a soft brush or a feather with which to brush bees from the frames, a jar of petroleum jelly, and a comb stand. This last, a very useful accessory, consists simply of two stout uprights on large feet. Spars are fastened to the uprights, which are spaced just wide enough apart to allow frames to be hung by the lugs. Its uses are obvious, for it is frequently necessary to move a frame from the hive while examination is in progress, and on such a stand it is perfectly safe from injury. Similar carriers are often fitted to the walls of each hive. The petroleum jelly is to brush on the lugs of the frames and the bars on which they hang in the hive. It serves a useful purpose, as it will prevent any possibility of the bees sealing them down with propolis. Just the merest trace should be smeared on with a brush or the finger.

Whenever possible, bees should be examined during fine, warm weather and while they are flying freely. The best time to open the hive is undoubtedly about midday. Then, if properly handled, the insects are more or less harmless. No attempt should ever be made to examine a hive on chilly days, during wet weather, or when there are cold winds. Even in warm weather it is wise to defer operations if there is thunder in the air.

The smoker must on no account be over-freely used. In this respect beginners are always apt to err. It is certainly inadvisable to attempt to open a hive without first subduing the occupants until one has acquired complete confidence and knows the temperament of one's stocks, but the less subduing bees receive the better. The effect of either smoke or carbolic cloth is simply to frighten the insects. As a result they immediately repair to the honey cells on the comb, and gorge themselves with food. In a state of repletion they are not at all likely to sting, for the simple reason that their abdomens are distended and they cannot readily curve them to drive their stings in perpendicularly.

Unless the brood nest itself is to be opened up, it should not be necessary to blow any smoke into the entrance. If the brood frames are to be inspected, one or two puffs can be blown in. Then, working from the side of the hive, remove the roof and all the quilts save the bottom one. Raise this gently at one corner, blow a little smoke over the bees, and then replace it for a moment or two, when it can be pulled off gently, at the same time smoking the bees down from the top bars.

If a carbolic cloth is preferred to a smoker, this should be drawn on as the last quilt is removed. This is done in the one operation. The cloth is held by two corners and allowed to hang down over the side of the hive farthest from the operator. The corners of the quilt are then gripped and the cloth pulled over the frames as the quilt is drawn towards the operator. The cloth must be left in place until the bees are quiet, generally a minute or two, and should they get at all restless at any time it must immediately be replaced.

If there is only one dummy board in the brood nest, operations must commence from that side. If there are two, it is of little consequence. The first step is to draw back or remove the dummy to give a little lateral space for withdrawing the frames. These must be taken out very carefully, so as to avoid knocking them against the hive side or rubbing the bees between them. A start should be made with the nearest one, which must be drawn sideways before being lifted from the hive. If the outside frame is broodless, it need merely be drawn to the hive side before proceeding with the next. It is always advisable to keep as many of the frames as possible covered up, especially when the bees are at all irritable. To do this two quilts are used. One is laid over the uninspected frames and rolled back as each one is removed. The other is drawn forward over those that have been examined as each one is replaced.

There is a right and a wrong way to hold and turn frames as they are examined, and the beginner must make himself thoroughly proficient in this by practising with an empty frame before attempting to handle full combs. Unless these are manipulated in the proper manner, accidents will almost certainly occur. The

48

frame should, of course, be lifted by the lugs in both hands, with the top bar uppermost. After the nearest side is examined raise the right hand until the top bar is perpendicular. The frame can then be swung round like the leaf of a book and the hand lowered. This will leave it with the bottom bar uppermost, and the other side of the comb can then be examined. To return it to its original position with the top bar uppermost before replacing in the hive, the movements are simply reversed.

Examination of the combs will reveal honey cells, larvæ, and eggs. The honey is stored above the brood, in cells immediately below the top bar. Those nearest the bar capped over with light-coloured wax are full of ripe honey. Immediately beneath them will be other cells, as yet uncapped, containing unripe honey. Still farther down on the comb will be the cells containing hatching brood. These have a darker capping of wax and pollen mixture. Any of obviously large diameter and with cappings which protrude notably are drone cells. In addition, there will be open cells containing larvæ, others with eggs, and nearest the entrance capped and uncapped cells of pollen. Queen cells which have already been described and are illustrated in one of the photographic reproductions are quite unmistakable. These are very much larger cone-shaped structures, which hang from the edges of the comb.

In the W.B.C. type of hive, which has the frames running at right angles to the entrance, honey, brood, and pollen are usually found on all combs.

Finding the Queen in a crowded hive is not an easy job for the beginner, though the expert will usually trace her in just a few minutes. In the early part of the season, she will most probably be found about the centre of the brood nest, but only providing quilts have been used to prevent it being flooded with light. As

D

soon as she is discovered on a frame it is well to return it to the hive without delay, for there is a chance that she may take wing. In this event there is only one possible course of procedure, and that is, to remain quite still and wait for her return.

To remove the bees from a frame a soft brush can be pressed into service, but this is not as a rule necessary, and should not be practised unless it is really essential. Another method is to scare them off by holding the frame firmly by one of the shoulders above the hive. The other hand is then sharply thumped on the one holding the frame. The bees will drop off and scurry down amongst the other frames. This is hardly a method the beginner is advised to adopt, though providing the frame is very firmly gripped, and just one sharp thump is employed, it is most effective. It must not, however, be adopted if there are queen cells on the frame.

In a really strong colony all ten frames should be well filled with honey, brood, and pollen cells, and to ensure that careful check is kept of the progress made by each stock, a hive card, on which all records relating to the colony may be noted at each examination, should be kept. This is best tacked inside the hive roof. Particulars to be noted include number of frames of brood, number of queen cells formed, age of Queen, date of swarming, date of installation of supers, date completed, etc. A complete record of the stock's activities and past performance will show at a glance its true strength. Each time inspection is made a close lookout must be kept for any signs of the diseases detailed in a later chapter. Nothing is more fatal to the prospects of an apiary than to allow disease of any kind to spread unchecked.

Just as soon as examination is completed, the dummy should be replaced or drawn up and the quilts laid on one by one. Before so doing, however, it may be necessary to smoke the bees down from the frame tops. By replacing the quilts one at a time any bees which may be caught are given a chance to escape.

Despite all precautions, it is certain that the bee keeper will be stung at some time or other. There are many so-called remedies for stings, but the first thing is to remove the sting itself, if it is left in the flesh. This must be done by scratching it out with the finger-nail or a knife. On no account attempt to pull it, as this will simply result in more poison being injected from the detached poison sac. A little ammonia or the domestic blue bag rubbed gently over the affected part will soon reduce the pain and swelling. Actually a bee will remove the sting itself if it is not frightened in any way. To stay perfectly still without trying to brush the insect off after having been stung is, however, beyond the powers of most people. The least fright and the insect will be away, leaving its sting behind it, for to withdraw it, it must circle round, to roll up the barbs.

CHAPTER
NINE
❋

THE HONEY
HARVEST

IT is a very common fallacy amongst beginners that
bees gather and store surplus honey the whole
summer through. Except in heather districts, where
there are really two distinct periods of activity, the
honey flow is confined to little more than six weeks,
commencing as a rule about the end of May. The
bee keeper's whole endeavour must be to have his stocks
as strong as possible at this time and to have all pre-
parations in the way of fitting out supers made some
time in advance. It is a fatal fault to be late in installing
them, for, as explained in another chapter, overcrowding
will encourage swarming.

The first point to be decided is whether to work with
sections for comb honey or with frames for extracted
honey. The ideal is, of course, to have a proportion of
both, and where honey is only required for home consumption, sections are as a rule preferred. Bee keepers
who make their hobby a profitable side-line have, however, several objections to working with sections. First
of all, it must be realised that a colony will produce
at least one-third and probably one-half more extracted
honey than it will comb sections. With frames, too,
the same combs can be used year after year, providing
uncapping and extracting are carefully carried out, thus
effecting a great saving in bee labour. Marketing of
extracted honey is also very much simpler, for there is
no real risk of breakages nor of deterioration. Bottled
honey, if properly stored, will keep in perfect condition
for years. Sections, on the other hand, must be sold
at once or they may granulate.

There is, of course, the point that sections fetch a rather higher price, but the difference is so slight as to be hardly worthy of consideration, in view of the great saving effected in other ways by working for extracted honey. The extracting outfit naturally adds to the initial expense of setting up an apiary, but it is not an unduly large item and will prove money well spent. In an apiary of any size the cost of foundation and the sections themselves will be an annual item just as large as the initial price of a good extractor and uncapping knives.

A start should be made with the preparation of section or frame racks several weeks in advance of the commencement of the honey flow. With the W.B.C. hive, twenty-one sections fill a rack in three rows of seven. These must all be fitted with full sheets of foundation and placed in the rack with a separator between each one. The separator is simply a very thin sheet of zinc or tin cut to the same shape as the sections. When placed between them, it prevents the bees from drawing out the cells beyond the edges of the boxwood. The long, slotted separators are easiest to handle. These will cover the face of three sections, and have bee ways cut in them to provide ready access. Short separators are exactly the size of one section. Wooden separators are seldom used nowadays, as they are always fragile, and unless very carefully handled breakages will be all too frequent. On no account must they be omitted, for if the cells are drawn out beyond the edges of the section it will be quite impossible to pack it for transit.

The sections are firmly held in the crate by means of boards and spring locks, as shown in one of the illustrations. Although perfectly secure they are therefore very easily removed.

There is another type of section rack which, though rather more expensive, has an advantage, in that it allows the bees the freest possible access. This is the W.B.C. hanging section super. In this the sections are hung in frames each containing three, just in the same manner as frames in the brood body. Although more expensive, there is no doubt that this type of super possesses decided advantages, especially with stocks which prove loath to enter the crates.

A FITTED SECTION CRATE

The sections are held quite firmly in position by means of boards and *spring* locks. They are therefore easily removed.

A shallow frame super is similar to the brood body in every respect, save that the frames are not so deep. Actually it is not essential to use shallow frames in the supers, though it is claimed by most experienced bee keepers that the bees take more readily to them, as they increase the storage accommodation of the hive more gradually than if a rack of standard frames is given.

It is possible to use ten narrow (⅞-inch) frames in the super as in the brood chamber, but a more usual method is to hang eight wide (1⅜-inch) frames. These are spaced by 2-inch instead of 1½-inch metal ends. Drone foundation is used and the result is wider and deeper cells, for which less wax is required in proportion to the honey stored in cells. Incidentally, no pollen will be stored in these big cells, which is a great advantage. These frames must be fitted with full sheets of foundation, or, better still, empty combs, for then the bees are saved the labour of drawing out the cells. Until they are required, the racks must be carefully stored and covered up safely. On no account must they be allowed to get damp or dusty, while mice, and the very destructive wax moth which lays its eggs in the foundation, must also be guarded against.

The first signs of the commencement of the honey flow are to be seen in the whitened edges of the cells near the top bars of the brood chambers. This is caused by the bees drawing out the cells with fresh white wax for the accommodation of the new honey. This is the very latest date at which supers should be installed. With strong colonies they should, of course, have been in place some time beforehand to discourage swarming.

Installing the supers is actually quite a simple task. The bottom of the crate should first of all be smeared with petroleum jelly to prevent the bees gluing it down with propolis. Then the roof, lift, and quilts should all be removed, save for the bottom sheet. As this latter is peeled off, the carbolic cloth should be rolled over the frames to drive the bees down from the top bars, or alternatively a puff or two of smoke can be blown over the tops of the frames for the same purpose. As a rule, however, it is not necessary to use a smoker, and if it can be done without so much the better. Any

55

propolis or wax on the top bars should be scraped off with a paint scraper.

The super containing the eight shallow frames will fit exactly on top of the brood chamber, but before it is placed in position, the queen excluder must be laid on top of the brood frames, but below the cloth.

Section racks are very light, and it is quite possible to place them in position with one hand while withdrawing the carbolic cloth with the other. Another point, a queen excluder is not always essential where sections only are worked for. The Queen, as a rule, will not make use of them for egg laying, unless the weather proves unfavourable. The top portion of the hive under the quilts is naturally the warmest part of the hive, and in cold, wet spells she may ascend. All things considered, it is perhaps wiser to use an excluder, for though it offers a little hindrance to bees laden with honey, that is a small point as compared to the fact that the sections will be completely ruined if the Queen does choose to use them.

It is also claimed that wide frames, spaced two inches apart, as just described, are rarely used by the Queen, but here again the risk of dispensing with the excluder is hardly worth while.

Frame racks are both heavier and deeper than section racks, and it is not possible to hold them in one hand. The simplest thing for the beginner is to have an assistant at hand who can remove the cloth, while he himself places the rack in position.

Only one super should be added at a time, but others must be given long before the first one is completed. This is always essential, for the bees do not fill and cap the cells without delay. The honey is left uncovered for a week to ripen, and the storage room in one super will be exhausted long before the process of sealing is commenced. Just as soon as the first super is about half full a second one should be given, not above, but under the first. Bees can then work on this while the honey in the top rack is maturing. Further racks should be given as necessary, and always in advance of the bees' requirements, but when it is seen that the honey supply is failing, no more should be given, for the bees must then be encouraged to cap over all the surplus ready for removal.

In a good season and with strong stocks, the bees will fill a rack of shallow frames at an astonishing rate. A week will see the cells drawn out, and it is by no means uncommon to find them completely filled within a fortnight. To remove the rack, it should first be twisted gently to see that it is quite loose. If it has been well smeared with petroleum jelly, it should come cleanly away. If, on the other hand, propolis has been used to fix it down, a screwdriver must be pressed into service to lever it up. Thin pieces of wood can be inserted as each corner is raised.

A carbolic cloth should be drawn across the brood frames as the first super is removed. If when clear it is held over the cloth for a minute or two, the bees will be driven up into the sections or frames, when the new super can be slipped underneath it. Make sure that the two supers fit quite evenly, one on top of the other and, of course, on top of the brood chamber beneath. Finally, the carbolic cloth is drawn out and the hive closed up again.

When tiering, as the practice of adding supers is termed, is attempted without smoking the bees, the quilts should not be removed from the top of the first super, or the bees are likely to give trouble. It is certainly advisable to try to perform the operation without doing more than place the carbolic cloth above the brood nest, for the loss of even a few hours at the height of the honey flow is a serious matter, and bees which have been smoked will do little more work that day.

If, as sometimes happens, the bees prove very loath to work in section racks, they can be induced to do so by including in the rack one or two unfinished sections kept over from the previous season, especially if they happen to contain a little honey. Special frames are also available which hold six sections, and can be inserted in the body box quite early in the season. If these are later moved to the super, the bees will not be long in setting to work.

A SUPER CLEARER

Above is shown a bee escape, mounted on a board. The bee's only means of escape is through this, once the slide shown in front is closed.

Unless the honey is required as soon as possible, it is certainly best to leave shallow frames on the hive until the honey flow is over. It is not, however, advisable to leave sections on the hive after they are capped, as they will certainly be stained and soiled by the bees. In their case it is best to remove the racks and sort out the contents as soon as the central sections are sealed.

This is invariably done some time in advance of the capping of the side rows. All finished sections should then be removed, the unfinished ones placed in the centre, and the side rows filled in with fresh ones ready fitted with foundation.

It is quite permissible, too, to take off one or two empty combs or new frames, but this will involve the use of a super clearer. This is an ingenious appliance which has made a very simple task of what was once a most difficult operation, for with its aid the surplus honey can be removed without trying the patience of the bees. It consists simply of a wooden board which exactly fits the base of the super. In the centre is a bee escape, the principle of which is clearly shown in one of the illustrations. The bees descend through the round hole, and then pass through the very fine springs, to the brood body. Once out they cannot return, as they are unable to force their way through the springs from outside. The illustration shows a single bee escape, but double or quadruple escapes, i.e. with two or four pairs of spring outlets, can also be obtained and save time when clearing a crowded super.

To place a bee escape in position it is again an advantage to have an assistant, for then it is only necessary to lift the super a couple of inches above the brood body and slide in the board. Within twenty-four hours all the bees working in the supers will have descended through the escape, and the supers when removed will be quite empty. The operation is one which must be carried out with the utmost care, for if the bees are upset they are liable to break the cappings on the combs. For this reason smoke or the carbolic cloth should not be resorted to unless it proves absolutely essential.

Supers should always be removed as soon as the honey flow is over. If the job is delayed until the nights turn definitely chilly, the bees will begin to transfer the honey to the brood frame.　In most districts the end of July will see the main flow completed, when they should be taken off without delay. Experience may teach, however, that it is wiser to wait a little longer, and in heather districts, it will be well on into September before the flow ceases.　Just as soon

A BEE ESCAPE IN DETAIL

The bees descend through the round hole shown on the left and pass out through the fine springs shown in the cut-away section.

as the supers are removed, they should be moved indoors, where flying bees have no access to them.　If left about in the open the smell of the honey may incite bees to start robbing.　There is nothing more difficult to stop once it commences in real earnest.

CHAPTER
TEN
❋

SWARMING

ALTHOUGH everyone is aware of the honey bees'
habit of swarming, it is to most an inexplicable
mystery. At first sight it seems nothing short of
folly that the bees should suddenly abandon the hive
after labouring to lay up stores of honey and nectar,
and leave it to those which remain, and the young
brood hatching in the combs. But swarming is simply
their natural method of increase, and it is only in
keeping with the selfless nature of their lives that they
should be ready and willing to surrender the fruits of
their labours to the future generation, and themselves
seek new quarters in which to start afresh.

The principal cause of swarming is overcrowding
in the brood chamber, with consequent overheating in
the hive. The Queen, as previously mentioned, is
capable of laying eggs at the stupendous rate of 3,000
a day when honey is coming in freely. With a really
strong stock bringing in heavy loads of nectar and
pollen which fill a fair area on the combs in the
brood chamber, her sphere is materially reduced, and
there will soon be insufficient room for her laying
activities. Then it is that the swarming fever gains a
hold on the bees. Queen cells are built, and as soon
as these are capped over, first indications of swarming
will be noted. A strange restlessness will possess the
workers. A good proportion of them will go about
their duties as though nothing was in the air, but
others will not stray far from the hive. Instead, they
will be seen flying restlessly around the entrance. Then,
within a short time, a swarm can be expected, and

some time between the hours of ten and four on a warm sunny day a good half of the hive's inhabitants will stream from the entrance, taking with them their mother-Queen. As she joins them, they will gather round until she settles on a nearby branch or tree, when they cluster round her in a dense mass.

Then they are easily secured, and can at once be hived, but it is a job which must not be delayed, for even as they issue from the hive, scouts are despatched to seek for some new home, and this will be at some considerable distance from the parent hive. Should the taking of the swarm be delayed until the scouts return, they will at once fly off, heading in a straight line for the new home which they have never seen. To follow them once they have taken wing from the cluster is no easy job, for they seldom travel less than two or three miles to some hollow tree or dark, dry cranny.

There is one point which must be impressed at the outset, and that is, that swarming bees are practically harmless. Even a child need have no fear of handling them, for before leaving the hive they will have gorged themselves with honey, both to enable them to subsist themselves and to provide the wax necessary for comb building in their new home.

About the only indication that a swarm was imminent with the old skep hives was the presence of flying drones. Under modern conditions the bee keeper can tell almost to within a day when the exodus may be expected, and that some two weeks in advance, for as soon as queen cells are built, it is obvious the bees are thinking of a move. From the laying of the eggs to the sealing of the queen cell a period of nine days elapses. If careful note is kept in the hive register of queen cells built and the date on which the eggs are laid in them, there will be no trouble in determining very closely when the swarm may be expected. It does not of course follow that a swarm will issue with clockwork-like regularity as soon as the cells are capped. Swarming may be delayed or even abandoned altogether in the event of rain or stormy weather. In this case the bees will tear down the queen cells and destroy the royal nymphs. It is seldom, however, that swarming is abandoned after one attempt has been ruined by the weather. As a rule a fresh batch of cells will be constructed and the swarm will merely be delayed.

This is the worst thing which can happen. Early swarms, which issue in May from a strong colony, have ample time to build up combs and increase their strength before the main honey flow. The parent stock too will be in a good fettle for that all-important and all-too-short season. Late swarms, which issue in June and July, will not have time to rear sufficient brood to make a really strong colony, and will produce little or no surplus. It is, in fact, no uncommon thing for late swarms to be unable to store sufficient honey for their own supplies during the winter. Exception must, of course, be made in the case of heather-honey districts. There the honey flow is prolonged until well on into the autumn.

Whether swarming is an advantage or not depends entirely on the plans the bee keeper has in view. If additional stocks are required, a good early swarm will be a welcome sight, and providing fresh quarters are available in which to house them, one or two strong swarms are not to be regretted in a good season. To ensure against their being lost, however, steps should be taken to induce the swarm to cluster quite close to the hive. If there are no adjacent trees or suitable places for them to settle, it is a good plan to plant some boughs near the hive. Very often this will prevent them from clustering in difficult and sometimes inaccessible positions. If the bee keeper is at hand when the swarm issues they can also be induced to cluster by spraying some water over them with a garden syringe.

When a swarm clusters on a branch or bough where it is easily reached, it is a simple matter to secure it. All that need be done is to take a skep and holding it close under the swarm dislodge the bees into it with a vigorous shake. The skep should now be turned right way up on to a cloth spread on the ground, but it must be blocked up from the ground with a few stones or pieces of wood, and be shaded from sun, with another cloth. Within a very short time the bees will have clustered inside the skep. Then it must be picked up and carried gently to the position the new hive is to occupy. It must on no account be left for long in the wrong position, or the bees will start to mark the spot.

The evening is the best time to hive the bees. This is done in exactly the same way as detailed in an earlier chapter, simply shaking them on to cloth-covered boards running up to the alighting board of the hive.

When swarms cluster in thick hedges, on high trees, or other inaccessible positions, a different method of securing them must be adopted. A simple scheme when they cluster in a thick hedge is to obtain an assistant who will hold the skep just above or beside the bees, keeping the edge close to them in order that they may ascend into it. This they are induced to do by smoking them gently from below. Methods to meet the exigencies of the moment must be adopted when swarms settle on high trees. In some cases they may prove absolutely inaccessible. The only thing to do then is to wait and watch until they move and then endeavour to follow them. In other cases they can be secured by sawing off the branch on which they have clustered, but where this is done, great care must

SWARMS WHICH CLUSTER IN THIN HEDGES OR ON LOW
TREES ARE VERY EASILY SECURED. THEY ARE DISLODGED INTO
THE SKEP BY JERKING THE BRANCHES.

HIVING A SWARM. THIS MUST BE DONE IN THE EVENING, AND THE
BEES ARE SIMPLY THROWN ON TO CLOTH-COVERED BOARDS RUNNING
UP TO THE ALIGHTING BOARD OF THE HIVE.

be taken to see that they are not dislodged by an inadvertent shake. The same method as advised in the case of hedges, smoking them into a skep from below, can also be adopted where possible.

If the swarm should settle on a tree trunk, the best procedure is to brush them into the skep with a soft brush or feather, while swarms on the ground should be secured by placing the skep quite close to them but raised an inch or two on stones or pieces of wood. If some of the bees are induced to enter by brushing them with a feather the others will soon follow. In every case it is of the utmost importance to ensure that all the bees are secured, otherwise the Queen may be lost, in which case those in the skep will soon decamp.

The hive prepared for the swarm must of course be perfectly clean and fitted up with comb or comb foundation. With a strong swarm the full ten frames should be given from the outset. If the hive has previously been used, it is as well to scald and wash the inside with a solution of carbolic acid, but it must be allowed to air thoroughly for some time before the swarm is expected, as the smell of the disinfectant might result in the bees forsaking their new home. As already mentioned, all the quilts save one must be removed for twenty-four hours after hiving, as swarming bees generate a high temperature, and unless this precaution is taken, the heat may result in the wax foundation breaking down. There is a distinct possibility, unless steps are taken to prevent it, that an after-swarm or cast will issue from the parent hive headed by the first young Queen to hatch. This happens as a rule exactly nine days after the main swarm has issued. If left unchecked, and especially in the case of Carniolans, others may follow as the young Queens hatch, with the result that the parent stock is left useless. It is, however, quite a simple matter to

E

prevent casts. This may be done by going over the combs of the colony as soon as the bees have swarmed, and cutting out all the queen cells save one, which must be left to furnish the new Queen. This, needless to say, must be a large and promising one.

From the point of view of the honey harvest, swarming in most cases is definitely to be prevented, but it is too late to attempt to do so once the bees have already got the fever. Instead the bee keeper must set himself to take steps to prevent their natural desire. The first thing must obviously be to prevent over-crowding. The bees must always be given room over and above their requirements, both in the brood body and in the supers. Stocks which do not cover the full complement of frames should have additional ones given before they are actually required. In very urgent cases where the brood nest is already full, it is even permissible to remove one or two combs of brood and give them to a weaker colony, filling their place with old combs which have been extracted. Foundation does not always solve the problem in this event, for if honey is being brought in rapidly, over-congestion may result before the cells can be drawn out on the new frames. It is therefore wise whenever possible to keep one or two old frames in reserve.

Ventilation must also be carefully looked to or over-heating will result. This too will encourage swarming. The entrances should always be kept fully open during the summer, as should the ventilator in the floor board, while the body box can be raised from the floor board by means of half-inch blocks placed under the corners. It is also a good plan to tilt or raise the roof slightly and to shade it from direct sunshine.

There is no doubt that the presence of numerous drones also encourages swarming. By using only worker cell foundation in full sheets the production of drones can be limited to a minimum, and in cases where even this is not sufficient, the unwanted drone cells can be cut out of the frames.

There is, of course, one method which might at first sight seem the obvious way of preventing swarming, and that is, to cut out all queen cells as they are formed, for the bees will not leave the hive unless they are assured that a new Queen is hatching to head the parent stock. Actually, however, this is not a good method to adopt. Even though the combs are examined weekly and all queen cells are cut away, the bees are liable to build fresh ones. The very fact that they have built the cells is an indication that the swarming fever already possesses them. Moreover, all the work undertaken is simply labour thrown away, if even one cell is missed. In nine cases out of ten, it results merely in delaying and not preventing a swarm, and during the extra time taken in the building of the new cells the bees about to swarm will be listless and doing no work, a serious matter during the honey flow.

Another method of swarm prevention must be mentioned, though the work entailed is perhaps rather more than the beginner may care to undertake. It is, however, a very sure means of preventing swarming taking place at all. It is necessary to have at hand a second brood body, which is fitted over the one already in position, but with a queen excluder between them. First of all, however, all the combs save one containing brood in the original brood nest should have queen cells cut out and be transferred to this upper chamber. The one left must contain both unsealed brood and eggs and must carry the Queen; the place of the combs transferred is then filled in with empty combs or frames fitted with wired foundation. In twenty-one days all the brood will have been hatched from the combs in the upper chamber, and since the Queen has not access to these, they will then be used for the storage of honey.

The Queen has, of course, no inclination to leave the hive, as she has, after the transference of the brood body, a complete new set of combs on which to set to work.

Clipping the Queen's wings also provides a simple solution to the problem. Although this may seem an extremely difficult and delicate task to the beginner, it is not over-troublesome if done with care. The best time to attempt it is fairly early in the spring, when the Queen is most easily found. Once she is discovered on a frame, draw it out and leave the carbolic cloth over the brood nest. If the comb is now hung on a stand the Queen can be picked up gently by the wings with the finger and thumb of one hand. She is then taken very gently by the thorax between the finger and thumb of the other hand and the tips of the large wings are clipped away with a pair of small, sharp scissors. The great thing is to handle her as gently as possible and to avoid at all costs exerting any pressure on the abdomen.

There is just one drawback to this method, and that is, if the swarm emerge unnoticed, the Queen, being unable to fly, will fall to the ground and may be lost; it is really essential to be at hand when the swarm emerges, so that she may be picked up. After all queen cells have been cut out, she may be returned to the hive. The swarm, on finding her missing, will return of their own accord.

Chapter eleven

FEEDING, WINTERING, AND SPRING MANAGEMENT

L EFT completely to themselves, bees in any normal season are fully able to provide themselves with sufficient food. But when kept on modern lines they are not given even a chance to do so. By far the greater proportion of their hard-earned stores is taken from them, and at a time which gives them little chance to gather more to make good the loss.

One of the main objects of feeding is, therefore, to keep the bees alive during the winter months, and where honey is not present in sufficient quantity, the food must be given quite early in the autumn. For if food is scarce, breeding will soon cease, and in extreme cases, the bees themselves may even destroy unhatched eggs and larvæ in an endeavour to limit their numbers according to the stores available. This is fatal, for apart from the fact that the stock will by the spring be so weak that it will be of little value, there is a grave risk of small colonies succumbing during the winter. During the cold months, the stock hangs in a cluster in the centre of the hive, and it is only the combined heat of the mass which enables them to resist cold. A small colony cannot generate the necessary temperature, and if it falls below a certain limit, all will succumb. Every endeavour must therefore be made to encourage breeding and so increase the numerical strength of the colony, until well on into the autumn. This can only be done by feeding.

A stock of normal strength requires approximately twenty-five to thirty pounds of honey if it is to winter safely. The first step must therefore be to estimate

69

the amount of food already in the hive. This should be done after the supers are removed at the end of the honey flow. It is, of course, only possible to estimate

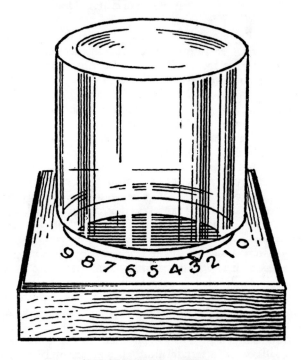

THE PERFECTION FEEDER

This is ideal for stimulative feeding in spring or summer, as the rate of feeding can be instantly regulated by turning the arrow to the required number.

roughly how much the bees have stored for their own requirements, but with a little practice it should not be found a difficult task. Honey cells, which are invariably at the top of the frame, are quite easily recognised from the brood cells by their white cappings, and the proportion of honey in each can be calculated on the basis that one side of a frame completely filled with honey weighs approximately three pounds. If, on inspection, the stocks' stores prove to fall short of the required amount, feeding must be started to encourage the Queen to maintain her laying activities. At the same time, it must be borne in mind that overfeeding is almost, if not equally, as bad a fault as underfeeding, for if provided with unlimited supplies of food, the workers may fill so many cells on the frames, that the Queen's activities will be sadly restricted.

For late summer stimulation the graduation bottle type of feeder is best. This can be set to feed both slowly and rapidly. It is not as a rule necessary to give a great deal just after the supers are removed, for all that is wanted is just sufficient to encourage uninterrupted breeding. The syrup can be made at home, but only the very finest grades of pure cane sugar must be used. Poorer qualities are distinctly injurious to the bees, as is syrup which has been even slightly burned in the cooking. Allow half a pint of water to every pound of sugar. Boil the water and pour it straight on to the sugar, then stir until dissolved. Do not continue to heat on the stove, owing to the risk of burning the sugar. As soon as the syrup has cooled it should be poured off into bottles, which are then tightly stoppered and stored in a warm, dry cupboard until they are required.

The feeder is placed directly over the brood frames and must be warmly covered up. This is a rule which applies to feeding at any season of the year, otherwise the bees will forsake it. As a rule a quarter to half a pint of syrup per night is sufficient, but inspection will soon show just how much should be given, and the feeder can be set accordingly. The point to remember is that none of this syrup is intended to be stored in the combs. At the first sign of the bees doing so, supplies should be cut off.

About September a start should be made with autumn feeding. Sometimes a thicker syrup is used for this—up to two and a half pounds of sugar per quart of water and with the addition of a tablespoonful of vinegar to prevent crystallisation. The aim now is to give syrup to the bees just as quickly as they can take it down and store it in the combs. If the graduated feeder is used, all holes may be exposed, but really the rapid feeder is to be preferred at this season. This autumn feeding must on no account be delayed or continued until too late in the season, for if some of the syrup stored in the cells is not sealed down, it will most probably ferment, and when consumed by the bees will give rise to dysentery. Deficiencies in the bees' stores must be made good in time to allow them to cap the syrup before cold weather puts a stop to further activities. This means that syrup feeding should never be prolonged beyond the end of September.

The safest rule is to give one pound of syrup for every pound shortage below the normal store of thirty pounds of honey in the hive. Wherever possible, the feeder should be refilled every evening, for the more quickly the bees can be induced to carry the syrup to the combs the better. It is a good plan, too, to warm it slightly before filling the feeder. It must on no account, however, be overheated; if it is just lukewarm, that is quite sufficient.

There is one danger which is always present when feeding is in progress after the cessation of the honey flow, and that is, robbing. As already remarked, this is very easily started, and once the bees get the fever it is extremely difficult to stop. Precautions should therefore be taken just as soon as feeding commences, by contracting the entrances of the hives. Where this is done, it is unlikely that robber bees will be able to gain entrance to the hive.

To the novice, feeding on such a scale may seem not only expensive but to involve a good deal of trouble. In view of the fact that next season's honey harvest depends entirely on the strength of the winter stock, it must, however, be accounted labour worth while and money well spent. The more so, when it is considered that the cost of the sugar is trifling in comparison with the market value of honey. It is, too, a matter of only a few minutes to refill feeders each evening, and all the syrup required can easily be made at one time and stored for future use.

In addition to syrup feeding, it is always a sound plan to give each stock a cake of candy before closing it down for the winter. Even hives which have ample stores of honey should be given this "iron" ration, for if the bees do not require it, it will not be eaten and can be removed in the spring. It is, on the other hand, a fairly certain precaution against starvation, which is responsible for the loss of more colonies than is readily imagined.

The cluster hanging in the centre of the hive may easily be reduced in size by cold snaps killing off many of the outer bees, with the result that they are unable to reach some of the stored honey, for there is no isolated movement in the hive during the winter. If the bees were to attempt to move singly for even a short distance over the combs they would be numbed and killed by cold. The candy placed on the frames directly above them ensures that they remain in actual contact with food, and it can be renewed if necessary throughout the winter.

Candy is not so easily made as syrup, and in view

73

of the fact that it can be very cheaply purchased, it is hardly worth while to endeavour to do so. However, here is a recipe for those who prefer to make it at home. Place in a clean aluminium or copper pan three pounds of best white sugar and half a teaspoonful of cream of tartar. Add half a pint of water and warm very slowly, stirring constantly, until all the sugar is dissolved. During this preliminary melting the pan is best placed beside rather than over the source of heat, to reduce risk of burning. Then heat more rapidly until the liquid boils and continue to boil steadily for two minutes. Remove pan and stand almost to its rim in cold water till the syrup looks cloudy. Then pour into saucers or other shallow moulds. When quite cold the candy should be firm but moist.

Candy need not be given until the hive is closed down for the winter, and if purchased in a glass-topped box can quite easily be inspected without danger of chilling the bees. A hole about three inches square must, however, be cut in the centre of the bottom sheet. This should never be disturbed throughout the cold months, for at the first approach of severe weather, the bees themselves will seal it down with propolis to ensure against fatal draughts. The candy is placed directly over the hole. Warm quilts are then heaped over the sheet. Any odd material can be used for this purpose. Fairly thick pieces of felt are ideal, but squares cut from old carpeting are almost equally serviceable if sufficient are used. It is wellnigh impossible to have too many coverings and on top of all should come a loosely packed chaff cushion. This can be stitched up out of sacking or any similar material. If there are any empty combs, the brood nest itself should be contracted by means of dummy boards, after removing the unnecessary frames. The entrance should then be contracted to about four inches, and the roof secured by driving a stake into the ground on one side of the hive and passing over the roof a rope on which a couple of bricks or a heavy stone are hung. This simple precaution is always well worth while, for storms may possibly upset the hives or lift the roofs. The roofs themselves must, of course, be perfectly waterproof, and if any doubts are entertained in this respect, a piece of canvas should be tacked over and painted. Damp entering the hive will have very serious effects on the health of the colony.

Examination during the winter is only necessary to ensure that the bees have sufficient food. Just one or two quick peeps on fine warm days will be quite enough. Even then it should be done as quickly as possible, and just a corner of the quilts lifted. One glance, if the candy is contained in a glass-covered box, will be sufficient. If the cake has simply been placed on top of the frames, its size must be gauged by feeling it through the sheet with the fingers.

It is never worth while to winter weak stocks, for apart from the danger of their dying out during the winter and the fact that they are not likely to show a profitable return in the following season, they always consume a greater proportion of food than do strong colonies, in an endeavour to maintain the necessary heat. This is another very frequent cause of dysentery. About the end of August or early in September, all stocks that cover less than six or seven frames should be united, as explained in a later chapter. The better of the two Queens is kept to head the colony.

SPRING STIMULATION

Spring feeding, like summer feeding, which takes place immediately after the supers are removed, is not given with a view to replenishing the bees' stores in the combs, but to encourage brood rearing, in order that the colony may be at its best for the commencement of the honey flow. It is, to borrow horticultural parlance, a simple method of " forcing " brood rearing, for so long as food supplies are coming in regularly the Queen will continue to lay. It is obvious, therefore, that only just sufficient to keep the bees busily employed in handling food but not storing it should be given, and a start must be made at least six weeks before the honey flow opens, for it is just over four weeks from the laying of the eggs before the young workers hatched therefrom are able to take up their chief task of nectar and pollen gathering. Stimulation is always with a view to having strong young workers ready to leave the hive just as soon as there is scope for their labours.

The graduated feeder is the type employed and the same syrup as used for summer and early autumn feeding. As a rule, it is quite sufficient if two or three holes are opened, and that only during the night. Even then, a careful watch must be kept to see that none of the syrup is stored in the combs. Every available cell must be left for the Queen with the exception of small reserve stores of honey in the side combs. In the event of a spell of unfavourable weather which prevents the bees from flying freely, it may be necessary to increase supplies a little, but it is seldom advisable to give more than a quarter to half a pound per day.

In most districts, March or April should see the commencement of spring stimulation, and it should cease as soon as the bees forsake the feeder. This they will do whenever honey is coming in in sufficient quantity. It can then be removed.

If there is an abundance of sealed food in the hive, syrup will not be required, for the bees can easily be induced to handle food by bruising about two inches of the cappings every week with a knife passed down between the combs. Pollen is, of course, essential in addition to honey, and if it is obviously scarce it is a sound plan to dredge some pea-flour over some clean shavings in a small box, which must, however, be protected from rain. Alternatively, some of the vacant cells on one of the outer frames can be filled with it and the frame moved in close to the brood combs. Wheat flour can be used if pea flour is not available. Another method is to add pea flour to the candy. This is done while the candy is cooling. The pea flour is sprinkled in a little at a time and stirred thoroughly. It is necessary to allow a quarter of a pound of flour to every three pounds of sugar.

SPRING CLEANING

About the end of March a very careful examination should be made and the hives thoroughly cleaned. This is a task which must never be neglected, for nothing is more likely to give rise to disease than a dirty hive. A fine warm day must be chosen for the task, and if a spare hive is available, it will prove a great help, for the frames of the first colony examined can be put into it as they are inspected. As soon as examination is completed the new hive is then moved to the site occupied by the old one, which after cleaning will do for the second colony. The work is continued in the same manner until all have been dealt with. As each old hive is cleared it should be scraped with a painter's scraper and the floor board thoroughly washed with strong soda and water. The top bars of the frames should also be scraped clean.

Examination of each frame must be careful and thorough. Keep a close watch for disease and defective combs. The latter, if broodless, should be removed. If they do contain brood, move them towards the sides of the hive with a view to removal later on. It is also a good plan, whether the combs are defective or not, to remove at least two each year, choosing the worst of the old combs and giving in their place foundation-fitted frames. In this way the combs are completely renewed every fifth year. It is true that they will last much longer, but the size of the cells is reduced each year through the cocoons spun by the larvæ adhering to the walls. In time this reduction in size will make for less vigorous stocks.

Careful note should be taken of the amount of food and brood in each hive, and should the colony turn out to be weak, the brood must either be contracted with a division board or united with another stock, the better of the two Queens being retained to head the colony. It is always as well to make certain of the presence of a fertile Queen. If by any chance she has succumbed during the winter and the stock is left queenless, the only course is again to unite with a colony which has a laying Queen, for at this season of the year Queens can rarely be obtained.

Very great care must be exercised to see that the bees are not upset more than is inevitable when hives are opened in early spring. The less smoke used the better, for if rudely disturbed, it is quite possible that the bees may ball their Queen, quite a number clustering round her in a tightly packed mass. Very often they maintain their hold until she is either squeezed to death or at least seriously injured. There are several ways of dealing with balling, one of the simplest being that of immediately closing down the hive and leaving it for a week or so until the bees have quietened down. There is no certainty, however, that the bees will release the Queen when closed down. An alternative is to lift the ball and drop it into a basin of water. So great will be the bees' alarm that they will at once break apart. The Queen can then be rescued and caged on the comb for a few days. Smoking the bees to compel them to relinquish their hold is not always a good plan, for when this is done, she is quite frequently stung and killed. The same is liable to happen if any attempt is made to break up the ball with the fingers.

Spring examination, though it must be thorough, should be completed as quickly as possible, and the brood nest warmly covered up the moment it is finished, or unsealed brood may be badly chilled. The frames in particular should not be exposed for one moment longer than is necessary.

CHAPTER TWELVE ❀ UNITING AND INCREASING STOCKS

IF there is one thing the novice must always bear in mind, it is that weak colonies are of no use whatsoever. The only course, where stocks have become weak, is to unite them. Two stocks, which alone could not possibly earn a profit, will, if united, show a handsome return. In the autumn too, it is essential to unite any which do not cover at least seven frames, for smaller colonies would, in all probability, perish through their numbers being insufficient to maintain the necessary temperature in the hive.

There are other instances where uniting is necessary. Colonies, found to be queenless in early spring, must at once be united to a stock headed by a fertile Queen, and any nuclei used during the summer for Queen rearing must either be joined up or united to stronger colonies in the autumn.

It is not possible to unite bees of different colonies without first of all taking precautions, or they will certainly fight, and since bees distinguish members of their own family from strangers by their scent, the simplest and safest method of procedure is to give both colonies the same scent. Uniting fluid, which can be sprayed over the bees before they meet, can be purchased from any specialist, or the same object may be achieved by dusting them over with flour.

Where two stocks have to be united, a spare hive is a great help, though not essential. The first step is to move together the two hives containing bees. This should be done two feet at a time, till they are side by side, or with the spare hive close between them. Leave them like this for a couple of days. The frames are then transferred to the new hive, or to one of the old ones if no other is available. Alternate those from each hive and dust all the bees adhering to them with flour before they are placed in position. It is unlikely that there will be room for all the frames, in which case those which are not covered with bees (almost invariably the outside frames) are rejected. Any bees remaining in the old hive, after the frames are moved, should be thoroughly dusted and shaken on to the frames before replacing the sheets, quilts, and brood.

Only one Queen will be required, so the better should be selected and the other caught and destroyed or used elsewhere. The one that is to be kept must be caged on a comb for forty-eight hours until the bees have settled down. Then she can be removed. Special cages can be purchased for this purpose and, when the Queen has been placed inside, the cage is pressed into the comb (see next chapter).

Uniting must never be attempted during the day-time. Instead, it must be done in the evening, after all the flying bees have returned to the hive, for any which entered the hive after the union had taken place would certainly be killed, as they would not have the same distinctive scent of flour or uniting fluid.

TO DRIVE THE BEES DOWN
FROM THE TOP BARS OF THE
BROOD FRAMES A CORNER OF
THE QUILT IS LIFTED AND A
FEW PUFFS OF SMOKE BLOWN
OVER THEM.

THE METHOD OF REMOVING A SUPER IS SHOWN ABOVE.   AS
IT IS LIFTED OFF, THE CARBOLIC CLOTH IS DRAWN OVER THE
BROOD NEST.

ABOVE ARE SHOWN QUEEN CELLS ON A SECTION OF COMB. THEY ARE
EASILY RECOGNISED BY THEIR SIZE AND ARE USUALLY FOUND HANG-
ING FROM THE EDGES OF THE COMB.

THE COMPONENT PARTS OF A MODERN BAR FRAME HIVE ARE SHOWN
ABOVE. ON THE LEFT IS A RACK OF SECTIONS. CENTRE IS THE BROOD
NEST, WITH ONE FRAME WITHDRAWN, AND ON THE RIGHT A SUPER
FITTED WITH SHALLOW FRAMES.

Procedure when uniting a queenless stock to one headed by a fertile mother is identical. If the frames of the queenless stock do not contain fertile brood, the bees can be brushed or shaken into a skep or swarm-box, and after being dusted, can be hived in the same manner as swarms, throwing them on to a cloth-covered hiving board leading up to the entrance.

Weak stocks can also be strengthened by giving them a swarm from another hive, but, unless one is certain that the swarm is headed by a fertile Queen, it is wise to place a piece of excluder zinc over the entrance while the swarm is being hived. If a virgin Queen heading an after swarm is allowed to enter, it is more than likely that the fertile Queen will be the one to die. Procedure, when uniting nuclei in the autumn, is identically the same as that for uniting stocks. Where, as often happens, bees have been driven from a skep or procured to bring weak colonies up to strength in the autumn, it is advisable to take additional precautions against fighting, and again an empty hive will prove a valuable aid. This should be placed quite close to the existing stock. The driven bees can then be hived in it either on five empty combs or on five combs taken from the old stock and from which the bees have been brushed. After an interval of three or four days, they can then be united in the normal manner.

INCREASING STOCKS

Swarming is the bees' natural manner of increasing, but heavy swarms and a good honey surplus cannot be had in the same season. It is possible, however, by means of artificial swarms, to increase colonies without unduly prejudicing the honey harvest, providing all the stocks which are swarmed are really strong.

Actually, artificial swarming offers many advantages, for it prevents the loss of time inevitable when stocks

F

are preparing to throw a natural swarm. Instead of colonies buzzing about restlessly for several days before the swarm actually issues, they can usually be induced to start work in fresh quarters without any delay. Another point, where artificial swarming is practised, the bee keeper is assured of early swarms, and as is explained in a previous chapter, it is only early swarms that are really worth while.

The bee keeper too is enabled to ensure that any increase is effected from the strongest colonies headed by the best Queens. With really strong colonies it is permissible to split them into two. To do this, one frame of brood, together with the Queen and the ad-hering bees, should be removed from the old hive and placed in the centre of the brood box of a fresh hive. Then, six frames with foundation are placed, three on either side of this brood comb, and the brood box contracted with dummy boards. When the quilts and roof are replaced, the old colony should be moved to a new location and the new hive placed on its site. Providing this is done about noon, on a warm day, when the bees are flying freely, and the old hive is moved at least six feet away, all bees on the wing will return to the new hive, and form a good colony. If possible, a fertile Queen should be introduced to the parent stock, as this will avoid any delay in brood rearing. Where a Queen is not available, however, the bees can be safely left to rear one for themselves.

An even better method than halving the strength of one colony is to make a third out of two. One method of so doing is to take five frames, well filled with brood, from the first colony. Shake off all the adhering bees and place the frames in an empty hive; both new and old hives are then filled up with old comb or foundation-fitted frames. The new hive, containing only brood, is then moved to the site occupied by another strong

82

colony, this latter being shifted at least six feet away from the position it formerly occupied. Again the flying bees will return to the new hive. Wherever possible, a fertile Queen should be introduced without delay, but, if this cannot be done, the bees will rear one for themselves.

Where there are several colonies, none notably stronger than the rest, this method can be adapted to as many as need be. One comb of brood without bees may be taken out of each of five hives and then the new hive, so formed, should be placed in the position of a sixth well-stocked hive so that flying bees may be gathered to nurse the brood. If any bees are left on the brood combs from the other hives there will be fighting between the different stocks. The great point, in every case, is to make sure that artificial swarms are made on fine warm days, when the bees are flying freely, otherwise failure is something more than a possibility. It is a great help, too, to have some fertile Queens on hand, as this prevents any delay in brood rearing.

All newly introduced Queen bees must be caged for twenty-four hours (see next chapter).

QUEEN
REARING

THE whole success of any hive centres around the Queen. Without a prolific breeder which produces a good working strain, honey production will never rise above mediocre levels. Moreover, it is only young Queens which are capable of producing really strong stocks. As a rule, Queens are at their best during their second season. From that time on, they decline steadily, and if good yields are to be obtained, every hive should be re-queened just as soon as the Queen has passed her prime.

There is a fairly widespread belief that Queen rearing is a difficult and troublesome task. But in actual fact, procedure is perfectly simple and straightforward. So much so, that there is no reason why every bee keeper should not raise his own Queens.

In an apiary of any size, one or two spare Queens should always be on hand, for apart from their being required to replace old and exhausted sisters, there is always a possibility of accident. Very often, Queens are killed by being crushed when manipulating the frames, and there is, too, the possibility of their being balled by the bees. Yet another frequent cause of queenlessness is the inability of some virgin Queens to find their own hive when they return from their mating flight. Should they, by chance, enter a strange hive, immediate execution is their fate.

It is a very easy matter to ascertain when a colony is queenless, for as soon as the bees realise their loss, they are thrown into a state of great excitement, and will be seen running up and down and searching frantically for their missing mother, both inside the hive and out. The bees will continue their search until they give up hope and commence to build queen cells. But sometimes this is done so late that there are no larvæ in the hive young enough to produce good Queens. When they can be supplied with a fertile Queen or even a queen cell, as soon as the loss of the old one is noted, they will continue working steadily and without loss of time.

Where Queens are to be reared, nucleus hives must be prepared during the winter or in early spring. These are simply small hives large enough to take three or four frames. Where spare hives are available they may be converted by contracting them with dummy boards, and it is even possible, where a good many Queens are wanted, to accommodate three nuclei in the one standard hive. This is done by dividing the hive into three with dummies deep enough to touch the floor of the hive and so divide it into entirely separate compartments. Then additional entrances must be made for two of the stocks through the back and side walls of both brood box and outer casing and some provision made to prevent the bees flying into the space between these on their way in and out. If, later, the hive is used to house colonies, these entrances can be closed by plugs of wood or cork.

There are many methods of Queen rearing adopted, but the easiest is that of obtaining Queens from a freshly swarmed stock. Before leaving the hive, the swarm will have built queen cells over young larvæ, and since they leave nothing to chance, several of these will have been formed in the hive. A day or two after the swarm has issued, the combs can be examined and split up into three nuclei, making sure that each has at least one good big queen cell. Small ones should be removed. One of the nuclei can be housed on the site of the old hive. The others, which must be placed in new positions, should be given rather more bees to make up for the percentage of the flying bees which will return to the old stand. Within about a week, the young Queens will hatch out, but they must be kept in the nucleus hives until fertilised. Then they may be utilised as required. As soon as eggs are found in a nucleus, it is safe to assume that the Queen is fertilised.

Each nucleus hive must have, in addition to queen cells, comb containing brood, pollen, and honey, and on account of the small number of bees contained therein, it is essential to keep them very warmly wrapped up. Two frames of brood are sufficient.

Another method which does away with the difficulties of introducing a strange Queen is to give queenless colonies a ripe queen cell. Where this is to be done, the Queens which are to be got rid of should be removed at least two days in advance. The cells are then simply cut out and inserted between the brood combs. One essential with this method is to see that any queen cells given are ripe or within a few days of hatching. It is quite easy to tell when cells are ripe, even though note is not kept of the exact date when they are built, as the bees always remove some of the wax from the points of those which are about to hatch to enable the young Queen to make her way out more easily. In so doing, the wax is roughened.

The cells must be handled very carefully, and the smoker should be used to drive the bees from the frame before they are cut out. Thumping them off will almost certainly damage the larvæ. If, as is most probable, the cells are hanging from the edge of the comb, they should be cut out together with a piece of the comb above them. They are inserted between two combs of the hive's brood box and attached by turning down the piece of comb and pressing it on the frame

86

tops. The cells must be very carefully handled, for the least shaking or jarring may have disastrous results, as will exposure to a cool temperature.

A day or two later, examination must be made to see if the bees have accepted the cell. There is quite a chance they may have destroyed it and commenced to build queen cells on their own. These, however, must be cut out unless it is fairly certain that the larvæ placed in them were not more than three days old, for older larvæ will produce only laying workers. In such cases another ripe queen cell should be given. This method can also be adopted with nucleus hives.

Where cells from a swarmed stock are not available, the bees must be induced to raise them. This, too, is quite easily done. A frame of foundation or empty comb should be placed in the centre of the brood nest of the hive selected and examined after an interval of three days. If eggs are found, the Queen must then be removed and utilised elsewhere. In addition, all combs containing unsealed larvæ should be taken and given to other colonies, first of all shaking off the bees. These should be replaced with empty combs or foundation-fitted frames. The new comb containing eggs must now be withdrawn, and a strip, 1 inch wide, cut out of it, along its whole length midway between the top and bottom bars, making sure that there is a row of cells containing eggs along the top edge. From these, the bees will raise queen cells, which can be removed and given to queenless stocks or nuclei as described. When enough cells have been obtained, the cell-raising stock can be united or built up with frames of hatching brood, and given a fertile Queen.

As a rule, it is advisable to feed nuclei with thin syrup. The entrances must be kept contracted to guard against robbing, for, on account of their small numbers, the bees cannot well defend their hive against marauders. As already remarked, it is essential to keep them very warmly covered up.

To obtain good stocks, it is, of course, essential to see that queen cells are taken only from hives headed by the best Queens, and the points to watch, when determining the merits of several, are, prolificacy, especially in early spring, tractability, and the working qualities of the offspring. It is the drone which influences the disposition of the stock, and this too should be selected, so far as possible. This can be done by restricting drone production in the case of poor stocks, either by cutting out drone cells or by use of a drone trap. If the chosen stock does not produce drones freely, they can soon be induced to do so by giving them one or two frames fitted with drone foundation.

Certain precautions must be observed when introducing a new Queen, or she is almost certain to be killed, since she has not the same scent as the colony. First of all, it is necessary to make absolutely certain that the stock to which she is to be introduced is queenless. This assured, a frame should be withdrawn from the hive and the Queen placed in a cage. If the type known as a pipe-cover cage is used, a piece of card is slipped under the cage to prevent her escape. Card and cage are then placed upon the comb, the card withdrawn, and the cage pressed or screwed into the comb as far as the midrib. Some open honey cells must be enclosed by the cage, and it may be necessary to uncap a few. It is also essential to see that the Queen is not damaged in any way when pressing down the cage. The frame is now placed in the hive and the bees left severely alone for twenty-four hours. Then the Queen can be released. If the bees appear to accept her, she can safely be left at liberty. But if there is the least sign of their pulling her about, she must be caged again for another twenty-four hours.

88

There are many other types of queen cages, including some with holes filled with candy; in this case the bees themselves liberate the Queen by eating away the candy. All, however, work on the same principle of denying the bees access to the Queen until she has acquired the scent of the colony.

It is also possible to introduce a Queen without caging her, if it is done at night, when the Queen is hungry, and she is allowed to run into the hive from above. To make sure that she is ready to take food, the Queen should be placed in a matchbox for at least thirty minutes before her introduction and kept warm during that time. The hive should then be opened, the bees driven down from the top bars with a puff of smoke and the Queen allowed to run on to the frames from the box. This method has proved very successful, but with choice Queens it is really advisable to use a cage and so avoid any risk of loss.

CHAPTER FOURTEEN

EXTRACTING AND
HANDLING HONEY

WITH the aid of modern appliances extracting honey is a fairly simple task. But due care must be exercised and the various operations gone about in just the right manner, or there will be a grave risk of damaging the combs. It is a job which should never be attempted out of doors, for as a rule it is in progress at the close of the honey flow, just when the bees are most inclined to rob. Even when it is done indoors the first essential is to make sure that flying bees have not access, for if one or two find such ample stores unguarded, it is more than likely they will quickly lead others to the spot.

Before ever a start is made, the various appliances necessary must be got ready. Uncapping knives should be carefully cleaned and heated to prevent the honey sticking to the blade. When only a few combs have to be dealt with, this is usually done by having at hand a jug of hot water or a saucepan which can be kept hot on a stove. The knife is dipped into the water at frequent intervals as the work proceeds. If a good number of frames are to be uncapped, it is a decided convenience to work with two knives. One can then be left heating while the other is in use. The cutting edges must of course be just as sharp as it is possible to have them.

A large dish, a tray, or a clean board on which the cappings may fall is also required.

Now take a frame, and holding it by one of the lugs rest the other in the tray. The frame needs to be tilted so that the capping falls cleanly away as it is cut, and endeavour should be made to slice it off in the one cut, working from the bottom upwards. As soon as one side is uncapped, clean or change the knife and proceed in the same manner with the other. To become expert at uncapping requires practice, but with care the job can be managed quite successfully even by a novice.

As soon as the comb is uncapped it should be placed in one cage of the extractor and a start made on another of approximately the same size and weight, as this helps to balance the machine. It is always a good plan, too, to warm the extractor before use, as the honey is more fluid if kept reasonably warm and therefore the more easily extracted. The best plan is to wash it out thoroughly with boiling water and to work in a heated room.

The extractor must, of course, be firmly secured to the floor and the handle revolving the cages turned slowly for a start. Speed should be worked up gradually and at no time must it be more than just sufficient to throw the honey out. If the cages are revolved at great speed the combs may be cracked or even forced out of the frames. Only one side of the comb is extracted at a time. As soon as this first side is completed the frames are lifted out and reversed, unless the extractor itself is fitted with a reversing gear. The process is then repeated.

In the case of very heavy or soft new combs, it is a sound plan to reverse them when the first side is only partially emptied and return to it again after the second side is cleared. It is true that this procedure entails more work, but it prevents breakages which are something more than a possibility in both cases.

Just as soon as extracting is completed or the extractor requires emptying, the honey should be run into a ripener. It must then remain in a warm room for a few days. All the thin honey will rise to the top, together with most of the air bubbles. Thin honey

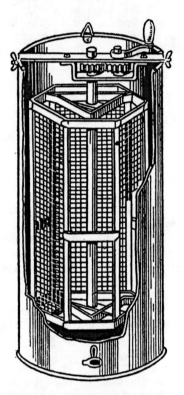

AN EXTRACTOR FOR THE SMALL BEE KEEPER

This machine, which deals with three frames at the one time, is of thoroughly practical design and will give first-class service.

should not be bottled for sale or consumption. By far the best plan is to feed it back to the bees, bottling only the best grade. Separation of the two is quite simple, as the honey is drawn off by means of the honey valve in the bottom of the ripener.

Combs which are extracted at the height of the honey flow can immediately be given back to the bees to refill. If, as is more probable, extracting is left to the end of the season, they must still be given back for about a week to be cleaned up in readiness for storing. As they are finished the frames should be put back in the super rack, which should be replaced on the hive in the evening to avoid exciting the bees. First of all, however, the super clearer must be placed above the brood box. When the quilts and roof are in position, open the slide at the side of the clearer. The bees will then make their way up into the supers and carry down all the honey left in the combs, leaving the cells perfectly clean and dry. If the slide is closed after about a week's time they will descend through the escape, leaving the supers clear. They can then be removed and stored away carefully for the winter in a cool, dry shed or room. It is a capital plan to wrap each frame in clean paper to guard against wax moth, while a close look-out will also need to be kept for mice, which can prove most destructive.

As already mentioned, heather honey, because of its much greater density, cannot be extracted from the combs, but must be separated by means of a honey-press. Small presses can now be purchased quite cheaply, or with very small quantities a potato masher can be used, in which case the honey is forced through clean cheese-cloth. As a rule, however, where only one or two hives are kept they are worked exclusively for sections.

When removed from the hive sections must be carefully scraped to clean off propolis or pollen stains. But the greatest care must be exercised not to damage the comb. A sharp knife is the best tool for the purpose, and as soon as they are all cleaned they should be re-

93

placed in the crates right way up, i.e. exactly as they stood in the hive. This is an important point, for the honey will leak from any uncapped cells and spoil their appearance if they are upside down or flat. Until used or sold, sections must be very carefully stored in a

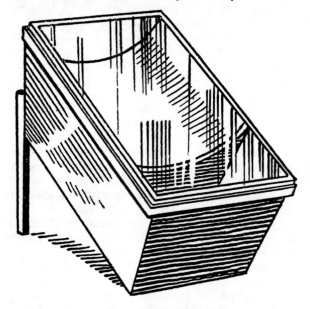

THE SOLAR WAX EXTRACTOR

The wax placed in the tray is melted by the sun's heat and runs down into the trough, leaving most of the impurities behind.

dry, warm room, and carefully covered over, to guard against dust or such unwelcome intruders as flies or mice. If left in cold or damp rooms the honey will probably go thin and leak through the cappings, or in a cold spell it may granulate; this, although a sign of purity, ruins its market value.

Bottled honey should be stored similarly in a dry, warm room, though it is of little consequence whether it granulates or not, as it is a simple matter to liquefy it. All that is necessary is to place the jar in warm water, resting it on a piece of wood to guard against the possibility of its being burnt. The heat of the water must not exceed 144 degrees Fahrenheit, or the flavour may be spoiled.

EXTRACTING WAX

Wax is a very valuable side-line, and not only the cappings, but also any scraps of comb which accumulate throughout the season, and odd clippings of foundations, should be carefully stored away to be extracted at the end of the year. Housewives find many uses for pure beeswax, and there is a ready market for it, as the firms manufacturing foundation find it no easy task to obtain sufficient supplies.

It is never advisable to mix old with new combs if the wax is to be sold, for the latter are always lighter in colour and fetch a higher price than the darker wax from old combs. The latter, too, usually contain a large amount of refuse and little wax. It is not necessary to go to the expense of purchasing wax extractors if only a small quantity has to be dealt with, as it is very easily rendered by boiling it out. The comb and cappings should first of all be broken up into small pieces and soaked in water for twenty-four hours. They are then placed in a clean canvas or cheese-cloth bag and boiled in a copper of soft water for two hours. A stone should first of all be put in the bag to weigh it down. The wax when melted will strain through the bag, and after the water is cool can be taken out in a solid cake. Most of the refuse will be left in the bag, but there will still be some dirty wax at the bottom of the cake. This should be scraped off. The clean wax can again be melted and run through a strainer into moulds, after which it should be allowed to cool slowly in order to avoid any risk of its cracking. The discoloured wax scraped off the bottom of the cake can again be boiled in a bag to remove more of the refuse. Even then, however, it will only yield wax of inferior quality.

The various types of wax extractors have already been described in Chapter Five. They are so simple to operate that nothing further need be said about them. There is no doubt that the steam-operated wax extractors will yield the purest wax, but, on the other hand, a good deal is left in the refuse, even though it is pressed through the strainer. The solar extractors do not render the wax so pure as do the steam types, but are the simplest of all to use.

Chapter fifteen

OTHER ITEMS OF IMPORTANCE

ROBBING

THE bee keeper who once experiences a severe epidemic of robbing is not likely to forget it, for unless prompt measures are taken to put a stop to it, a whole apiary may easily be demoralised and bees killed by the thousand. It is in the spring, when nectar is still scarce, and again just after the cessation of the honey flow, that the trouble is most likely to start, especially during the latter period. From then, right on into the autumn, the greatest care must be taken not to drop honey, syrup, or pieces of comb about the apiary, for just one or two drops of either honey or syrup spilled when manipulating the bees or while feeding, are quite sufficient to start piracy on a grand scale. Then, any weak stocks may be attacked by stronger ones, and even though they may resist the invaders stubbornly, they must eventually acknowledge defeat unless prompt aid is forthcoming.

The very presence of one or two weak stocks is in itself quite sufficient to start robbing, no matter how careful the bee keeper may be, and for this reason, too, it is essential to see that any which are below standard are united to bring them up to reasonable numerical strength.

There can be no mistaking the fact that robbing is in progress once it does start, even though on quite a small scale, for around the attacked hive there will be scenes of great agitation. Fierce fighting will take place on the alighting board, and the robbers will be seen hunting around the hive for an unguarded entrance;

from within will come an angry buzz as the defenders hurry to attack the invaders.

The obvious course when a hive is attacked is to contract the entrance without delay to one bee space. If this is done the robbers must make their way into the hive singly, and the defending bees have then a much better chance of dealing with them. A little carbolic solution or the carbolic cloth should then be placed on the alighting board; this may not prevent homing bees from entering the hive, but it will help to scare away the attackers. Where the robbers are extremely numerous it is also a good plan to spray them with water containing just a little carbolic solution, using an ordinary garden syringe for the purpose.

A piece of glass placed on end on the alighting board and resting against the wall of the hive at the entrance will still further confound the invaders in their efforts to gain entrance to the hive.

It is by no means uncommon for all the measures outlined to fail. In that event there is only one thing to do, and that is, to close the attacked hive entirely. Provision must, however, be made for adequate ventilation. A sheet of perforated zinc or wire gauze through which the bees cannot possibly make their way should be used to cover over the entrance and the slides can then be opened to their fullest extent. There is a danger that the trouble may recommence as soon as the hive is reopened and a close look-out must be kept.

In extreme cases it may even be necessary to move the attacked hive altogether until the robbing fever has left the culprits. But, if this is done, it is always worth while to place an empty hive on the site occupied by the one moved. Otherwise the robbers will in all probability have a tilt at the next weakest stock.

It cannot be too strongly urged that with robbing prevention is definitely better than cure, and, in addition to uniting weak stocks and making sure that no food of any kind is dropped about the apiary, all hive entrances should be kept contracted during the season when it is most likely to start. Nucleus hives, which have been used for Queen rearing, are always a potential danger, for since they do not cover more than two or three frames, they are naturally easy prey for a full-strength robber colony. The sooner nuclei can be united, once they have served their purpose, the better, and the entrances should never be wider than one bee space.

Where hives have to be opened, this should be done in the evening when there are no flying bees liable to be attracted by the scent of honey. Even then, manipulations should be carried out at the greatest speed compatible with safety. Feeding, too, should always be done in the evening, and hives where feeders are operating must have the entrances contracted even more than those that do not require additional stores.

The importance of storing honey taken from the hives well out of reach of any flying bees has already been mentioned, but it is as well again to emphasise the fact that if just a few bees gain access to it, they will soon bring others in their wake, and having once acquired the taste for stolen fruit, they will not readily recommence legitimate foraging expeditions. Even though steps are taken as soon as any bees are noted to have gained access to stored honey, to prevent their return will be too late, for, when denied access to the rich, unguarded stores, they will promptly turn on their weaker brethren.

Driving Bees

Bees driven from skeps are extremely valuable for uniting with weak stocks in the autumn, providing it is ascertained that they themselves are in perfect health and the combs show no signs of foul brood. Quite frequently they can be obtained at very trifling cost or even for the driving, and the beginner is therefore well advised to familiarise himself with the method adopted to induce them to leave their skep.

First of all the bee keeper must provide himself with a bucket large enough to hold an upturned skep, another empty skep, and a set of driving irons. These last consist simply of two right-angled irons about 15 inches in length and a rather shorter skewer.

Having everything at hand, a start is made by subduing the bees; this is done by blowing a few puffs of smoke into the entrance of the skep. In a few moments the bees will have run up between the combs and gorged themselves with honey. Now, place the skep to be driven upside down in the bucket, and attach the empty skep to it, bottom downwards, by thrusting the skewer through the edges of both. The angle irons are then pushed into the sides of both skeps to hold the topmost one at an angle away from the inverted one. The two skeps touch only where held by the skewer.

Two points must be carefully watched when fitting the second skep into position. They should touch at the end of the central combs, which must run from the operator, who stands in front, towards the skewer, and not across. Having made sure that the skeps are quite securely fastened, a start can be made with the actual driving. This is done by beating with the open hands or a couple of sticks on the side of the bottom skep. The beating must be steady and regular, both hands striking the skep simultaneously, and while it is necessary to jar the combs slightly, they must on no account be broken or loosened. The bees will at once leave the combs and run up into the empty skep, and if careful watch is kept it should be possible to spot the Queen.

Any tendency they may show to boil over the sides of the bottom skep can be checked by smoking them lightly.

Driving, for preference, should always be carried out on a fine day, when the bees are flying freely. The actual operation is best performed in some sheltered corner, and a box or a spare skep should be placed on the stand occupied by the one being driven to receive the flying bees as they return. When the operation is completed the driven bees can again be housed on the old stand and the flying bees which have returned hived with them by shaking them out on a board running up to the entrance.

With practice the whole operation can be carried out in a very short space of time, and the bees can then be united with weak stocks or with other driven lots as detailed in Chapter Twelve. As previously mentioned, the combs of the skep must be carefully examined for any signs of foul brood. If it is even suspected, the bees are not worth having as a gift.

Transferring Bees from Skeps

Where stocks have of necessity been wintered in skeps, they can quite easily be transferred to beehives in the spring without loss of precious brood, providing the colony is reasonably strong. Some time in April, when the population of the skep is increasing rapidly, is the time to make a start, but before then a frame hive should have been prepared with foundation-fitted frames. The tops of the frames are covered with an American cloth quilt, having a 4-inch hole cut in the centre. This is placed on the stand formerly occupied by the skep, which is then stood on the cloth above the frames. The entrance to the skep itself is stopped up and sufficient lifts placed on the hive, to permit of the roof being used above the top of the skep. The bees are thus obliged to use the hive entrance and to travel up and down between the frames.

Providing the colony is headed by a good Queen, it will not be long before the skep itself becomes over-crowded, and the bees will then proceed to draw out the foundation in the frame hive. Within a short space of time they will be followed by the Queen, and breeding operations will be carried on below. As soon as this is confirmd by the finding of brood in one or two frames, a sheet of excluder zinc should be placed under the skep to prevent her returning. As a rule, this will be within ten days or a fortnight of placing the skep in position.

Within three weeks of the Queen leaving the skep, all worker brood contained therein will have hatched out. It may then be removed and replaced by a super. There is just a possibility that drones hatching in the skep may choke the excluder. To obviate this, it is wise to tilt the skep slightly for a few moments to allow the drones to escape. This need not be done more often than about once a week.

MOVING BEES

There are due precautions which must be observed if it proves necessary to move hives during the summer months, for although bees will fly as far as two miles in their ceaseless search for food and return to exactly the same spot from which they started, a large number of them will be lost if their hive is moved more than two feet during their absence or at night. Where only a short distance has to be covered, it is really best to make the move in several stages of two feet or less, or alterna- tively the hive can be moved at night and steps be taken to ensure that the flying bees mark the new site. This can be done by placing in front of the entrance some object which will interrupt the bees' flight. A few bushy twigs laid on the alighting board will serve well, as will a cloth hung over the front of the hive. Even then, however, it is practically certain that a fair number will return to the old site. These should be collected and duly returned to the hive.

Providing the hive is moved for a distance of at least two miles, the trouble does not arise, for the change of scenery will then ensure that the new site is duly marked. Long journeys of this nature must be made at night, and preferably after a fine day when the bees have been flying freely.

In the winter hives may be moved any distance with complete safety, providing it is not attempted just after the bees have made cleansing flights. If weather has kept them confined to the hive for two or three weeks beforehand, there is no fear of their failing to mark the site on emerging.

LAYING WORKERS

Laying workers are only found in a hive which has been queenless for some considerable time, and should not therefore be a common occurrence in a well-ordered apiary. Various reasons are advanced for their presence. The most probable of these is that they are a result of belated efforts by the workers to rear themselves a new Queen, larvæ selected for the special treatment necessary to obtain a perfect female being more than three days old. As they are quite incapable of mating with the drones, laying workers can only lay infertile eggs, and these will, of course, produce only drones.

Were a hive containing only laying workers in the

place of Queens to be left to its own devices, the stock would inevitably perish, as the workers would dwindle steadily as the drones increased. Where examination is regularly carried out, however, it should be a simple matter to diagnose their presence long before an excessive number of drones gives rise to suspicions, as the brood is distinct from that of a good Queen in the scattered manner in which it is laid, as compared with the solid, orderly brood produced by a perfect female.

Once discovered, laying workers must be destroyed, and there is no better method of so doing than that of introducing a young fertile Queen. It sometimes happens, however, that where laying workers have been operating for some little time, the bees themselves will refuse to accept a stranger Queen. If they cannot be induced to do so the simplest course is to break up the stock and unite it with other colonies. The laying workers will then disappear, for their presence will not be tolerated in a hive where there is a reigning monarch.

The great thing, however, to prevent the appearance of laying workers is simply to see that no hive is allowed to remain queenless for long.

METAMORPHOSIS OF BEES

The metamorphosis or transformation of the bee from egg to the perfect insect varies in the case of the Queen, the drone, and the worker. It is in the case of the Queen, curiously enough, that the process is most rapid, the perfect insect leaving the cell on the sixteenth day after the laying of the egg as against twenty-two days in the case of the worker and twenty-five for the drone. In the following table are given the times in days which have been compiled by experts for the various stages of development:

	Queen	Worker	Drone
Period of incubation of egg .	3	3	3
Period of feeding the larvæ .	6	6	6
Period of spinning of cocoons	1	2	3
Period of rest . . .	2	3	3
Transformation of larvæ into nymphs	1	1	1
Time in nymph state . .	3	7	9
Total . . .	16	22	25

As a rule young Queens will spend about five days in the hive before embarking on their nuptial flights. Workers and drones, on the other hand, seldom fly freely less than a fortnight after they have emerged from the cells as perfect insects.

CHAPTER SIXTEEN

A BEE KEEPER'S CALENDAR

JANUARY.—There is little to be done this month except to make sure that winter storms have not caused any damage, and to clear any dead bees from the entrances. This is easily done by pulling them out with a hooked wire. Any roofs which show signs of leaking should at once be repainted, and damp quilts replaced with dry.

FEBRUARY.—In the event of a mild day, about the beginning of the month, opportunity should be taken to examine the food supply. If the candy given when closing down for the winter is exhausted, place another cake over the cluster, but disturb the bees as little as possible.

MARCH.—About the middle of the month, spring stimulation will start in early districts. Feed with syrup, uncap honey in the combs, and supply artificial pollen as soon as signs of activity are noted. Entrances must be contracted to guard against robbing. In the event of an early season, spring-cleaning and examination can be carried out towards the end of the month. As a rule, however, it is April before a start can be made. In any case, spare hives should be thoroughly cleaned and scraped in readiness, quite early in the month, and frames wired and fitted with foundation wax.

APRIL.—Spring-cleaning and examination should be finished as early in the month as possible. Keep a careful watch for any signs of disease or queenlessness. Stocks which are found to be queenless should be united at once with colonies headed by fertile mothers. Weak stocks, too, should be united, and in every case the better Queen must be kept. If both are good, one can be used to form a nucleus. Stimulative feeding must be continued and every effort maintained to build up strong stocks in preparation for the honey flow.

MAY.—Main endeavour must still be directed towards building up strong colonies for the honey flow. Where fruit blossom is plentiful, honey will start to come in during the month and feeders will be forsaken. As soon as this commences, and danger of robbing is over, open the entrances to their full width. Supers, both sections and frames, must be prepared quite early in the month, and given just as soon as first signs of the honey flow are noted. It is only in early districts that they will be required so soon, and they must be warmly wrapped up if the bees are to be induced to enter them. Ventilation must be ample and sufficient room provided to guard against swarming. Very strong stocks can have the brood nest doubled. In other cases, it will suffice to take away one or two combs of brood and replace with empty comb or foundation. The brood removed can be given to weaker colonies. Queen rearing can now commence, nucleus hives being formed for this purpose.

JUNE.—Swarming, where steps have not been taken to prevent it, will be in full swing this month, and clean hives should be prepared in readiness for housing swarms some time during the month. Supers must be given in good time, and fresh ones added as soon as the first are approximately half full. Clover honey will be coming in during the month, and particular care must be exercised to see that ample storage accommodation is provided. During hot weather, ventilation must be abundant and hives should be shaded from direct sunshine. Where artificial swarms are to be made, this should be done some time during the month.

JULY.—Ventilation and shade must still be provided and supers supplied where necessary, but storing must not be encouraged if the honey flow is failing. As soon as signs of falling off are noted, the bees must be encouraged to cap all honey already in the hive.

Queen rearing must be completed by about the middle of the month in most seasons. After that time drones may be destroyed. Hives which are due for re-queening should have this done as soon as the young Queens from nuclei are fertilised.

Robbing is always most likely to start just at the end of the honey flow, and great care must be taken, when removing supers, not to drop honey, syrup, or pieces of broken comb about the apiary. In addition, contract the entrances of nuclei hives to one bee space. About the end of the month, stocks which are to be worked for heather honey can be moved to the moors.

AUGUST.—Except in heather districts, all supers should be removed this month, and the stocks examined to ascertain their strength. Unite weak ones, or add driven bees to build up their numbers. With extracting in full swing, even greater care must be taken to guard against robbing, and where stimulative feeding is necessary, entrances must be contracted. Care must be taken to see that all hives are provided with good Queens.

SEPTEMBER.—Early in the month, weak stocks can still be united and, where necessary, all colonies should be fed up rapidly for wintering. Where possible this is best completed before the end of the month.

Hive roofs must be carefully examined and rendered perfectly water-tight.

OCTOBER.—All hives should now be snugged down for the winter. Place a cake of candy over the clusters, wrap up warmly, and weight down the roof. As soon as danger of robbing is past, open the entrances to about four inches. But, if necessary, wire them, so that mice cannot gain entrance.

NOVEMBER AND DECEMBER.—The bees should not be disturbed during these months. All that is necessary is to remove dead bees from the entrances. In the event of a snowfall followed by bright sunshine, it is also advisable to shade the entrances, as the brilliant light may tempt some of the bees outdoors.

CHAPTER SEVENTEEN

❀

PESTS AND DISEASES

DISEASE is fatal to the bee keeper's prospects. Allowed to ravage unchecked, most of the ailments to which bees are subject will soon reduce thriving colonies to a mere shadow of their former selves, and, in not a few cases, will eventually result in the destruction of the entire stock. With modern frame-hives, however, there is no reason why disease should ever have a chance to make progress, for the bee keeper should be able to discover it in a very early stage and take corresponding steps to check and cure it. In every case, however, it must be realised that prevention is infinitely better than cure, and there is nothing which will do more to promote good health than following the advice given in preceding chapters and maintaining strong stocks headed by vigorous young Queens. Cleanliness, too, must be the bee keeper's watchword, and here he is only following out the example set by the bees themselves. Regular re-queening is quite essential, for if an old Queen is allowed to continue in the hive, the stock will dwindle as her laying powers decrease. Weak colonies, too, must always be united, for they are an open invitation to disease.

In view of the contagious nature of most bee ailments, it is too much to expect immunity from trouble. But so long as there are no weak stocks where disease can thrive, it is unlikely that serious problems will arise, providing, of course, the necessary steps are taken to check progress of any disease which may make an appearance.

The so-called Isle-of-Wight disease, which was held responsible for much loss in the early years of the century, has, on investigation, proved to be, not a single ailment, but several unallied troubles which happen to produce rather similar effects. The designation "Isle-of-Wight Disease" should therefore be dropped, as it is misleading.

The worst of this group of disorders is Acarine disease. This is caused by a mite which lives in the bee's breathing tubes and gradually blocks them up by sheer numbers, besides damaging them structurally, and sucking blood from the bee. Symptoms of this malady are readily noted even by novices, for flying bees are greatly weakened. These will be found crawling around outside the hive, unable to fly through the blockage of their breathing tubes, and within a very short time they will succumb.

Whole apiaries so attacked may be wiped out of existence unless prompt steps are taken. The disease is spread by infected bees who may enter clean hives on robbing expeditions, accidentally through hives being placed too close together, or be introduced by the bee keeper himself when he unites weak hives or adds driven bees to weak stocks.

Preventive measures are simply those already outlined. Young Queens only should be used, and stocks must be kept as vigorous as possible. In addition, every precaution should be taken to guard against robbing. If, despite all care, Acarine disease does make an appearance, what is known as the " Frow " treatment will usually effect a cure. This will kill the mites without injuring the bees. The treatment is given during the autumn or winter, and consists in placing under the frames a pad of felt, on which is sprinkled one half-drachm of a mixture made up as follows: nitro-benzine two parts, safrol oil one part, and petrol two parts (all parts by volume). The dose must be accurately measured and sprinkled all over the pad, and should be repeated every day for a week. The pad should then be allowed to remain for another three days before removal. Ample ventilation must be given during this period, and the entrances should be opened to their fullest extent, and guarded by means of perforated zinc with only a small gap in the centre. In addition, the floor-board of the hive must be scraped out every day. Hives, which are attacked during warm weather, should have their brood given to another stock, special care being taken to see that all bees are brushed off the combs. The diseased stock may then be fumigated in the same manner, but for only three days. Later on, when all danger of robbing is past, they can be treated for the full time.

All supers should be removed from the hive before fumigation is attempted, and any which are left in the hive for the bees themselves must on no account be extracted for consumption, as the fumes of the mixture are poisonous.

Nosema disease is another of the maladies which were formally classed as Isle-of-Wight disease. This, again, is caused by a parasite, but one that is not so fatal as the Acarine mite. There is, however, a danger even in this, for the infected stocks may continue for years, during which time they will act as breeding grounds for the mite. Such stocks are seldom profitable, and this fact alone should make the bee keeper suspicious. Symptoms may be confused with those of Acarine disease, but there is little crawling until the bees are about to die, when they leave the hive, lie on their backs and make increasingly feeble movements with their legs. Microscopic examination is the only certain means of diagnosis of this as of Acarine disease, and in all cases of doubt it is advisable to submit specimens to an expert.

There is no satisfactory treatment for this malady, though it is probable that introduction of a young and healthy Queen may do something to save the colony, if it is done in the earliest stages. Where the disease has gained a stronghold, there is nothing for it but to destroy the bees, though the brood need not be sacrificed. It can, instead, be safely given to healthy colonies, as it is not susceptible to the disease. It has been proved that the spread of this disease is greatly accelerated if the bees are allowed to collect water from stagnant pools. Any such, in the neighbourhood of the hives, should be rendered unacceptable by pouring a film of paraffin on the surface of the water, while fountains which are frequently filled with clean water should be set up in the apiary.

Foul Brood is an equally, if not more, serious disease, chiefly because of its extremely contagious nature. One or two diseased colonies which are left untreated may easily be the means of spreading it through an entire district, for as the colonies are weakened by disease, they are robbed by other stocks which carry away the germs and spread the complaint. The spores may also be carried by the bee keeper himself from hive to hive.

There are actually two distinct diseases known as Foul Brood, namely, European and American. The

" American " disease is, in fact, more common in this country than the " European " variety, and these names have no relation to the actual distribution of the diseases.

As indicated by their names, these diseases attack the larvæ. American Foul Brood is produced by a bacillus (Bacillus larvæ). A healthy larva, as it lies at the bottom of the cell in the form of a C, is plump and of pearly whiteness. When infected by American Foul Brood it assumes a flabby appearance and lies horizontally. Within a short time its colour changes from white to an unhealthy yellow, and eventually to a dark-brown coffee colour. Then decomposition sets in, and the result is a mass of a glue-like consistency and odour, with a ropy tendency. As a rule, the cells are capped over before the death of the larvæ, and the cappings assume a dark colour. They are also frequently perforated. If a match is inserted into one of these cells and then withdrawn, it will pull out the ropy matter in a thread. Finally, the matter dries and adheres to the cells in the form of a brown scale. The bees themselves will frequently break the cappings and pull out the diseased brood, but this they will not do after decomposition has set in. Instead, they frequently fill up the cells with honey, which covers the decaying matter, but contaminates the food supplied to other larvæ.

Where there is any known danger of American Foul Brood, precautionary measures should certainly be taken. These consist of placing on the floor-board of the hive two split balls of naphthalene. In addition Naphthol-Beta should be added to the syrups and candies used for feeding. It is, indeed, quite a good rule to use this medicated syrup consistently. The Naphthol-Beta can be purchased from any commercial bee keeper, and should be added to the syrup at the

H　　　　　　　　　　　　　　　　　　　　113·

rate of one teaspoonful to every three pounds of sugar. It is stirred into the syrup while it is still hot. Alternatively, Izal can be used in the same way, at the rate of eighteen drops to three pounds of sugar.

Where the disease does appear, despite all precautions, it can be treated in the early stages by means of formalin, though a good deal of work is entailed in so doing, as each infected cell must have a drop or two of the formalin solution inserted with a fountain-pen filler after the capping has been broken. The solution used is one part formalin to four parts of water. A sheet of cloth soaked in a solution of one part formalin to fourteen parts of water is then placed on the floor of the hive, and renewed every twenty-four hours for as often as is necessary. The hands and all appliances used must be disinfected immediately afterwards, to avoid the risk of carrying infection through the apiary.

Where the disease is already in an advanced stage, it is best to destroy the bees with sulphur, or to call in the services of an expert. To do the former, open out a hole, three or four inches in depth, beside the hive, and place in this a tin lid containing a heaped-up tablespoonful of sulphur. If a red coal is dropped into the sulphur and the hive is lifted off its floor-board and placed over the hole, all the bees will be killed in a minute or two and will fall into the hole, when they should be sprinkled with petrol and burnt.

All hive fittings which have come in contact with the disease, including frames, combs, sheet, and quilts, must also be burnt, and the hive should be thoroughly disinfected by burning it off with a painter's blow-lamp, and then painting it over with a strong disinfectant.

European Foul Brood is believed to be caused by Bacillus pluton, but other putrefactive bacilli may be present and these greatly affect the degree of unpleasant smell produced. Such names as Sour Brood and Stinking Foul Brood have been applied to distinguish these symptoms, but they appear to be of little importance. In European Foul Brood the larvæ die as a rule before the cells are capped, and are usually removed by the bees themselves. Really strong colonies are often able to throw off an attack, but in bad years, when honey is scarce, it may become serious. The ropiness so typical of American Foul Brood is absent in the case of European and the scales which form when the matter has dried are as a rule quite easily removed. The obvious precaution, with this disease, is to have all stocks as strong as possible and headed with young and vigorous Queens, for then they will be able to throw off attacks. When the disease does occur in spite of precautions, it is best either to destroy the stock or to call in an expert. Re-queening may result in an improvement, but there is grave danger that the disease may break out anew later on and, in the meantime, other stocks may have become infected.

Strictly speaking, Chilled Brood is not a disease, as it refers simply to larvæ that have died through exposure to cold. Quite frequently it is mistaken by the beginner for Foul Brood, but it is easy to differentiate between the two. In the first place larvæ which have been chilled assume a bluey grey colour, quite distinct from the brown of Foul Brood. Later they go black, but there is never the odour of Foul Brood, which at its mildest is sour-smelling, and there is no difficulty in removing the dead larvæ from the cells. This the bees will as often as not do themselves, especially if the cappings of the cells containing the dead brood are broken. This is not a difficult task, as the brood which has succumbed is usually found in patches, not scattered over the combs. It is weak colonies or those which have had their brood nests unduly extended which usually suffer from this trouble. A cold spell of weather may cause the bees to contract for warmth, with the result that some of the brood are left exposed and are soon chilled. The obvious precautions are therefore to see that the brood nest is kept warmly wrapped up, while discretion must always be used before giving additional frames until danger of a late cold snap is passed. Another point, hives should never be opened on cold, windy days, nor should frames containing brood be exposed for one moment longer than is necessary, no matter how warm the weather.

Dysentery should never appear in a well-ordered apiary, as it is due to improper food, long spells of confinement, and disturbance during the winter. When bees are excited to such an extent as to encourage them to consume an excessive amount of food, or when they are fed on syrup made with coarse grades of sugar, they are liable to eject their excrement inside the hive. One very frequent cause of trouble is feeding too late in the autumn, with the result that the syrup is not capped over before the bees cluster. If left uncapped, it will ferment, and dysentery is then more or less certain. In very bad cases, it is wise to move the bees to a new hive and to give them fresh combs. They should also be fed and kept as warm as possible without disturbing them in any way.

Paralysis, a very mysterious malady, is fortunately not very prevalent, for little is known of it. With this complaint bees that have a shiny or greasy appearance and very few hairs may be noticed on the alighting boards. Later on, they are afflicted with a peculiar tremulous motion, the hind legs become useless and the abdomen swollen. Re-queening will sometimes effect a cure, but it is not a certain remedy.

Pests are few and far between. Probably the most destructive are wax moths. These gain access to and deposit eggs in the combs. The resulting larvæ work great havoc, tunnelling through the combs and devour-

116

ing wax and pollen. So long as colonies are strong, however, there is little fear of their gaining entrance to the hive. Carelessly stored spare combs are the ones which are usually attacked. To guard against this, spare combs should be carefully packed in supers, placing in each one or two balls of naphthalene.

Mice can also prove very troublesome by making their way into the hives during the winter. If they do gain entrance, it is more than probable that the colonies will forsake the hives in the spring, and as a precaution it is well to see that the entrance is protected by a strip of perforated zinc or by wires stretched tightly across it.

Wasps can also cause trouble, by robbing in the autumn, but if jars of sweetened beer are placed close to the hives, these will usually attract them. In addition, nests when found should always be destroyed.

CHAPTER
EIGHTEEN
✻

MARKETING HONEY

BEE keepers with more than just one or two hives as a rule contrive to make their hobby a profitable one. There is no reason why it should not be so. To find a market for first-class honey attractively packed is not a difficult matter. On the other hand, inferior grades or badly packed samples are never easy to dispose of, nor do they show anything like the same return. There is little to worry about concerning the quality of British honey, providing unripe or thin samples are not mixed with the finest grades. It compares favourably with any from overseas. But, as already mentioned,

thin honey is good for only one purpose, and that is,
feeding back to the bees. That marketed should be a
good level article, with little or no variation in colour or
quality.

Sections, in particular, should always be very carefully graded, for the price realised by a consignment is
as often as not governed by the poorest samples and
not by the best. As soon as they are removed from the
hive, the boxwood should be cleaned by scraping, as
detailed in Chapter Fourteen. The next step is to grade
them, and this is done by weight and appearance.
Those to be sold as first quality must turn the scale at
at least fifteen ounces. In addition, they should be well
filled and sealed, and the cappings must be perfectly
clean. Other grades can be sorted out according to local
requirements, but it is never worth while to attempt to
market badly capped sections or samples which are not
joined to the boxwood on all four sides.

It is invariably worth while to glaze sections, though
extensive use is now made of such materials as cellophane. This answers well, and serves the same purpose
of protecting the delicate cappings from the atmosphere
and from flies and other insects. There is no doubt
that the appearance of the sections is greatly improved,
while the cost of glazing or cellophane wrappers is comparatively small. Glass squares, measuring $4\frac{1}{4}$ inches,
some lace paper, and some good paste will be required
for glazing. The glass square is held in position by
means of the lace paper, which is pasted to the sides of
the section and then turned over flat and pressed down
on the glass. Both sides are treated in the same manner.

It is always advisable to endeavour to dispose of
sections locally, as they are not the easiest of things to
pack safely. No matter how securely they are padded,
jolting may easily break the cappings.

Extracted honey is nowadays put up in a great variety

of containers, but chief preference is still shown for screw-stoppered glass jars. These can be obtained in all sizes, but in most cases it is advisable to use the 16-oz. The jars must be spotlessly clean before filling, for even a suggestion of dust will show all too clearly, and detract much from their appearance and value. The jars are fitted with a cork wad, which effectually prevents leakage, but before being bottled, the honey should be stood in a vessel of hot water. Any bubbles will then rise to the surface. Not only does this improve the appearance of the honey, it also helps to prevent it from granulating. If it has already granulated before bottling, it can quite easily be liquefied as detailed in Chapter Fourteen. It then only remains to label the jars.

Attractive labels are supplied by several bee specialists, on which the name and address of the producer or the association marketing the honey can be printed. This is always worth while with good-quality honey, in view of the possibility of repeat orders. It is also quite a good thing to paste on a small notice to the effect that granulation is a sign of purity, and a natural consequence if honey is stored in a cold place, for not a few people still suspect adulteration in granulated honey. It is always a good plan to wrap the jars. Immediately after labelling the jars should be wrapped in white paper, for nothing looks worse than soiled labels, and the work and cost entailed are trifling.

The importance of packing honey in an attractive manner cannot be overemphasised. The very finest qualities will not find a ready sale or yield reasonable profit if badly put up. Even imported honey of much inferior quality will find a more ready sale.

Where extracted honey is to be sent by rail, travelling boxes holding twelve jars in separate compartments can be obtained, or it is quite a simple matter to pack them securely in any stout wooden case. The bottom of the box should be bedded with two or three inches of wood wool, each jar should then be wrapped in corrugated paper bedded firmly in the wood wool, and any spaces filled in with the same material. A good thickness of packing should also be placed under the lid, and a label, "Honey, this side up," attached.

The small keeper with only a trifling surplus to dispose of is undoubtedly well advised to try to retail his honey locally. Shops and stores are seldom interested in just a few pounds. There is the point, too, that by selling retail, profits are augmented by at least 25 per cent. At the outset it may not be too easy to work up a connection, but providing only first-quality honey is sent out, sales should increase yearly, for there is no better recommendation in any sphere of business than a satisfied customer. Larger quantities are actually more easily disposed of. The bee keeper with several hundredweights to clear is often relieved of the trouble of bottling and labelling, many larger stores preferring to buy the honey in twenty-eight- or fifty-six-pound tins and to do their own bottling.

Prices naturally vary with the seasons, but with extracted honey there is no necessity to unload the season's surplus on a falling market. Properly stored, it will keep perfectly, and the trouble of storing is often amply repaid by increased prices in following years, which may not be of such a nature as to encourage heavy yields.

INDEX

A

Acarine disease, 110
— — symptoms of, 110
— — treatment of, 110
After-swarms, prevention of, 65
American foul brood, 112
— — — symptoms of, 113
— — — treatment with
formalin, 114
Appearance of Queen bees, 16
Artificial pollen, 76
— swarming, 81
Assembling frames, 27
Autumn feeding, 72
— syrup, recipe for, 72

B

Bait sections, 58
Bee escape, 59
— space, 25
Bees, brushing, 50
— driving from skeps, 99
— feeding, 69
— handling in spring, 77
— sexes of, 13
— social life of, 15
— subduing, 47
— thumping, 50
— transferring from skeps, 101
Bottled honey, labelling, 119
— — packing, 119
— — storing, 95
Brace comb, 25
Brood body, 23
— combs, renewal of, 77
— frames, dimensions of, 27
— — handling, 48

Brown or black bee, 37

C

Candy, 73
Carbolic cloth, 33
— — effect on bees, 46
— — use of, 47
Carniolan bee, 38
Casts, prevention of, 65
Caucasian bee, 39
Causes of swarming, 61
Chilled brood, 115
Combs, storing, 93
— uncapping, 90
Cyprian bee, 39

D

Dimensions of sections, 27
— of shallow frames, 27
Driving bees from skeps, 99
Drone bees, 17
— cocoons, period of rest, 105
— — — of spinning, 105
— foundation, 28
— larvæ, period of feeding, 105
— metamorphosis of, 105
— production, limiting, 66
Drones, encouraging production
of, 88
Dutch bee, 39
Dysentery, 116

E

Eggs, period of incubation, 105
Embedding wire, 30
European foul brood, 114
— — — symptoms of, 115
— — — treatment of, 115

Excluder zinc, 21
Extracted combs, cleaning, 93
— honey, boxing, 119
Extracting honey, 90
— soft combs, 91
— wax, 95
Extractors, warming, 91

F

Feeder, graduated type, 33
— rapid, 33
— — principle of, 33
Feeders, covering, 71
Feeding bees, 69
— swarms, 43
Fertilisation of Queen bees, 20
Fitments for hives, 22
Fitting section racks, 53
— shallow frame supers, 54
Folding sections, 27
Foul brood, 112
Foundation wax, 27
— — grades of, 29
— — inserting, 29
— — securing, 29
Frame blocks, 31
Frames, assembling, 27
— fitting with foundation, 27
— ready fitted, 31
— scraping, 77
— types of, 26
— wiring, 29
Frow treatment, 111
Fuel for smokers, 33

G

Glazing sections, 118
Gloves, 32
Grading sections, 118
Granulated honey, liquefying, 95

H

Halving strong stocks, 82
Handling bees, 45
— brood frames, 48
Hanging section supers, 53
Heather honey, pressing, 93
Hives, 22
— fitments of, 22
— nucleus, 85
— position for, 44
— scraping, 77
— snugging down for winter, 74
— woods for, 25
Hiving swarms, 42
Honey, amount required for
    winter food, 69
— estimating quantity in hive, 70
— extracting, 90
— extractors, 35
— presses, 35
— ripeners, 35
— ripening, 91
— separating, 92
— thin, uses for, 92
Hybrid bees, 38

I

Inaccessible swarms, 64
Increasing stocks, 81
Indications of swarming, 61
Inserting foundation wax, 29
Installing supers, 54
Introducing Queen bees, 88
— — — — without caging, 89
— — cells, 86
Isle-of-Wight disease, 110
Italian bee, 37

L

Laying workers, 103
Life of worker bee, 16
Longevity of Queen bees, 18

M

Marketing sections, 118
Medicated syrup, 113
Metal spacers, 27
Metamorphosis of bees, 104
Mice, 117
Moving stocks, 102

N

Nosema disease, 111
— — symptoms of, 112
Nuclei, feeding, 87
Nucleus hives, 85

P

Paralysis, 116
Pipe cover cages, 88
Prevention of swarming, 66
Prolificacy of Queen bees, 15
Propolis, 25

Q

Queen bees, appearance of, 16
— — balling of, 78
— — cocoons, period of rest, 105
— — — — of spinning, 105
— — introducing, 88
— — — without caging, 89
— — larvæ, period of feeding, 105
— — longevity of ,18
— — metamorphosis of, 105
— — prolificacy of, 15
— — raising, 18
— — wing clipping, 68
— cells, encouraging production of, 87
— — handling, 86
— — introducing,86
— — removing, 67
Queenlessness, ascertaining, 84

Queenlessness, reasons for, 84
Queen rearing, 84
— — from swarmed stocks, 85
— selecting, 88

R

Recipe for autumn syrup, 72
— — summer syrup, 71
Removing supers, 59
Renewal of brood combs, 77
Ripening honey, 91
Robbing, 97
— causes of, 97
— prevention of, 98

S

Sections, dimensions of, 27
— fitting with foundation wax, 29
— folding, 27
— glazing, 118
— grading, 118
— marketing, 118
— racks, fitting, 53
— scraping, 93
— storing, 94
Section supers, hanging, 53
Securing foundation wax, 29
Separators, 53
Sexes of bees, 15
Shallow frames, dimensions of, 27
— — spacing, 55
— — storing, 55
— — supers, fitting, 54
Skeps, disadvantages of, 22
Smoke, use of, 47
Smokers, 32
— fuel for, 33
— use of, 47
Social life of bees, 15
Solar extractors, 36

Index

Spacing shallow frames, 55
Spring cleaning, 77
— stimulation, 75
Starting an apiary, 40
— — with stocks, 41
— — with swarms, 41
Stings, 51
— treatment of, 51
Stocks, increasing, 81
— — from any number, 83
— moving, 102
— strong, halving, 82
— three from two, 82
— uniting, 79
Storing shallow frames, 55
Subduing bees, 47
Summer stimulation, 71
— syrup for, 71
Super clearers, 59
Supers, installing, 54
— removing, 59
— tiering, 56
Swarming, 61
— artificial, 81
— causes of, 61
— indications of, 61
— prevention of, 66
Swarms, behaviour of, 62
— clustering of, 62
— feeding, 43
— hiving, 42
— inaccessible, 64
— inducing to cluster, 63
— in thick hedges, 64
— on ground, 65
— on tree trunks, 65
— preparing hives for, 65
— taking, 64
— timing of, 62
Syrup, preparing, 70, 71, 113
— storing, 71

T
Taking swarms, 64
Tasks of worker bee, 16
Thumping bees, 50
Tiering supers, 56
Timing of swarms, 62
Transferring bees from skeps, 101
Treatment of stings, 51

U
Uncapping combs, 90
— knives, heating, 90
Uniting driven bees, 81
— nuclei, 81
— precautions necessary, 79
— queenless stock, 81
— stocks and swarms, 81

V
Veils, 32

W
Wasps, 117
Wax extracting, 95
— extractors, 35
— moth, 116
— production, 28
W. B. C. hive, 23
Western red cedar wood, 25
Winter examination, 75
— food, honey required for, 69
Wires, embedding, 30
Woods for hives, 25
Worker bee, life of, 16
— — metamorphosis of, 105
— — tasks of, 16
— cocoons, period of rest, 105
— — — of spinning, 105
— larvæ, period of feeding, 105

Printed in the United States
82274LV00003B/99

9 781406 798036